ORGANIC REACTION MECHANISMS

Selected Problems and Solutions

William C. Groutas

Wichita State University

John Wiley & Sons, Inc.

New York • Chichester • Weinheim • Brisbane • Singapore • Toronto

To order books or for customer service call 1-800-CALL-WILEY (225-5945).

ISBN 0-471-28251-0

Printed in the United States of America

10 9 8 7 6 5 4 3 2

Printed and bound by Bradford & Bigelow, Inc.

To Susan, Mark, Christopher, Mom

And the Memory of Dad

PREFACE

The primary goal of this book is to use organic reaction mechanisms as a means of facilitating the mastery and understanding of the fundamental principles of organic chemistry, while at the same time sharpening a student's reasoning ability and critical thinking. This is achieved through the judicious selection and use of a large number of problems selected from the chemical literature. Each question is meant to illustrate one or more fundamental principles of mechanistic organic chemistry. The level of difficulty and suitability of the questions in this book have been tested by including them in exams and/or as homework assignments in the undergraduate organic chemistry course (Part A), or the first year graduate-level course in organic chemistry (Part B).

Special emphasis has been placed on the organization of the book. Part A contains questions geared toward students taking the sophomore-level organic chemistry course. The questions and principles illustrated thereof, are organized in the same sequence as they are normally discussed in a standard textbook of organic chemistry. A series of minireviews that summarize and reinforce fundamental principles that underlie a particular set of related problems have been included at the beginning of each set of questions. Thus, Part A can serve as a supplement to a standard textbook used in the first year organic chemistry course, and is intended to meet an existing need, since only a token number of end-of-chapter mechanism questions is included in most textbooks of organic chemistry.

The questions included in Part B are suitable for students in an honors course in organic chemistry, and beginning graduate students in chemistry, medicinal chemistry, biochemistry, and related disciplines. A limited number of applied problems has been included (Part C) to demonstrate how a knowledge of basic organic reaction mechanisms can be used to understand problems related to everyday life.

I am deeply indebted to Professor Richard A. Bunce (Oklahoma State University) for his diligence in reviewing the manuscript, and his many valuable comments. I am also grateful to Jennifer Yee for her editorial assistance, guidance and constant encouragement throughout the preparation of the workbook. Any errors that may have crept in are the sole responsibility of the author.

TABLE OF CONTENTS

ORGANIC
REACTION
MECHANISMS

Selected Problems
and Solutions

GLOSSARY

p-Toluenesulfonic acid (p-TSA)

N,N-Dimethyl formamide (DMF)

Dimethyl sulfoxide (DMSO)

Triethylamine (TEA)

4-Dimethylamino pyridine (DMAP)

1,4-Diazabicyclo[2.2.0] octane (DABCO)

1,8-Diazabicyclo[5.4.0] undecen-7-ene (DBU)

Dicyclohexylcarbodiimide (DCC)

Benzyl (Bzl)

A NOTE ON WRITING MECHANISMS

The majority of organic reactions can be viewed as Lewis acid/Lewis base reactions. Recall that a Lewis base is any substance that can donate a pair of non-bonded or pi electrons to form a covalent bond, and a Lewis acid is any substance that can accept a pair of electrons to form a covalent bond (see Minireview 1 for a general discussion of Lewis structures, Lewis acids and bases, and Lewis acid/Lewis base reactions).

In writing mechanisms, the following general approach should be followed:

(a) *add a sufficient number of non-bonded electron pairs on any heteroatoms* (atoms other than carbon, such as O, N, S, etc.) *to complete their octets*, since chemical structures are customarily drawn without the non-bonded electron pairs shown. By doing so, you will immediately identify the atom(s) in a given reactant that, in principle, can donate a pair of electrons to a Lewis acid. With just a little practice this will turn out to be a trivial task and, furthermore, in many cases you'll only be needing to add non-bonded electron pairs to the atoms that are directly involved in the reaction.

<u>EXAMPLES</u>

(b) *Identify the reactant that functions as a Lewis base, and the reactant that functions as a Lewis acid in a given reaction.* The reactant that acts as a Lewis acid typically has an atom that is electron-deficient, namely, it may have a positive charge, or a partial positive charge ($\delta+$) if it is bonded to one or more electronegative atoms. Examples illustrating these ideas are shown below.

EXAMPLE 1

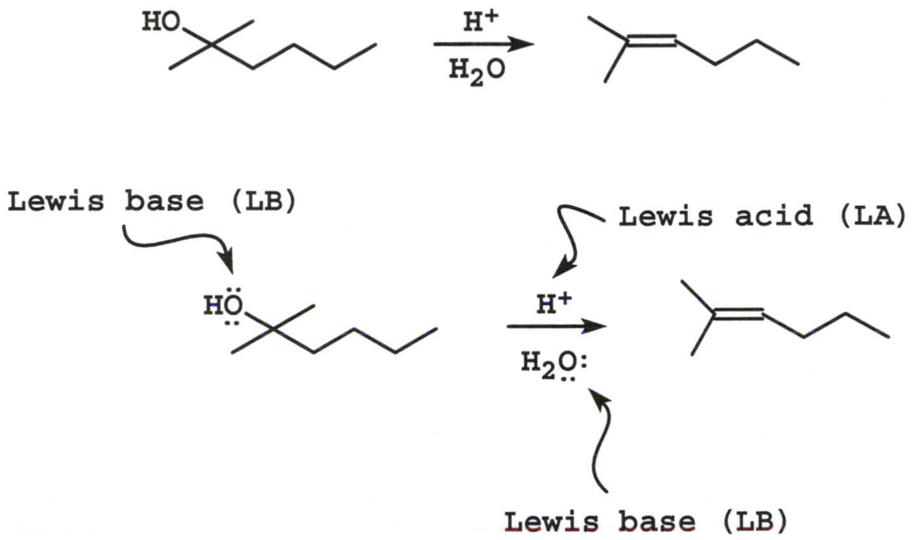

EXAMPLE 2

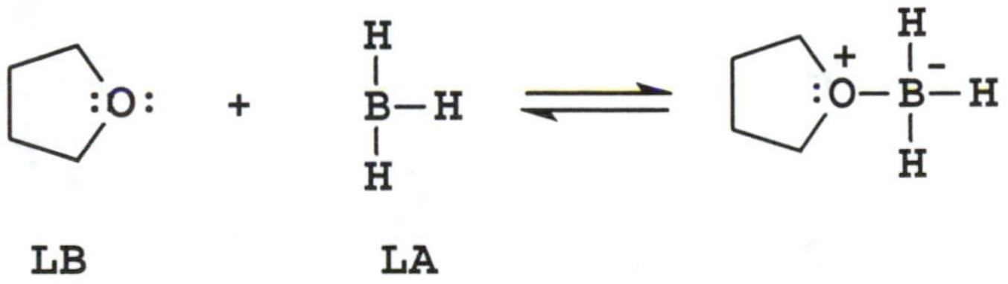

NOTE In general, compounds of group IIIA elements (B, Al, etc.) typically have an incomplete octet of electrons and invariably function as Lewis acids. Examples include boron trifluoride (BF_3), aluminum chloride ($AlCl_3$), boron tribromide (BBr_3), etc.

EXAMPLE 3

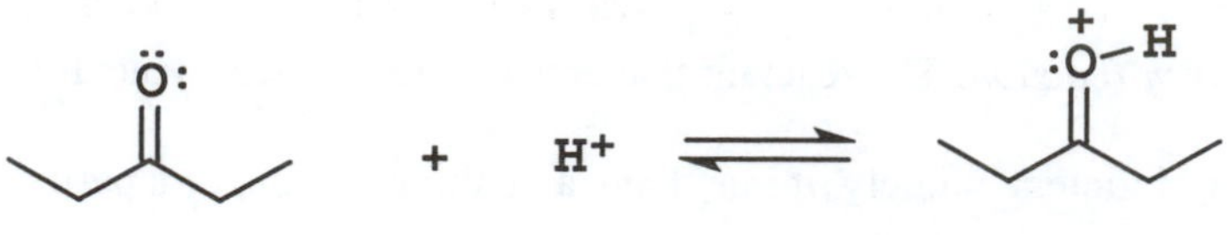

LB LA

EXAMPLE 4

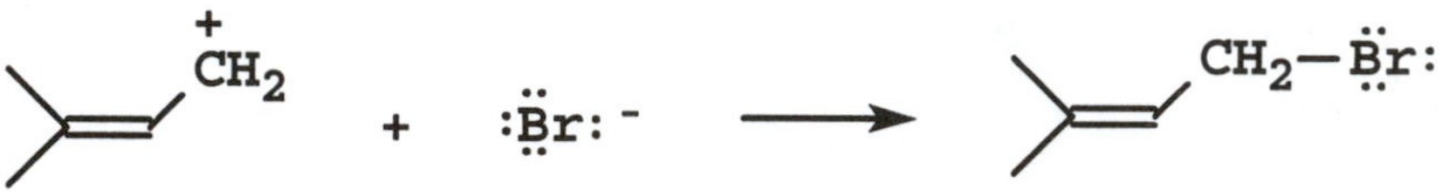

LA LB

EXAMPLE 5

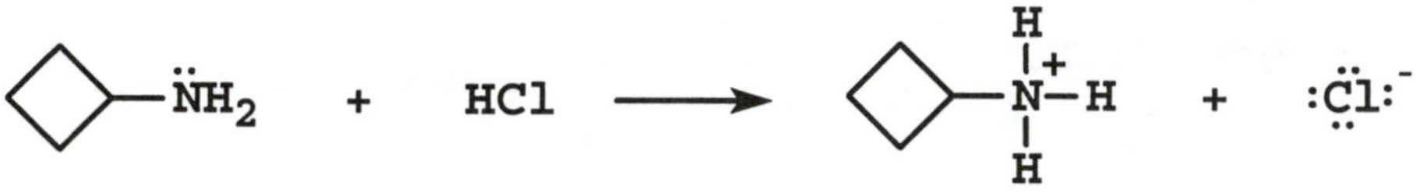

It is helpful to remember that strong mineral acids (HCl, HBr, HNO_3, etc.) or strong organic acids (for example, RSO_3H, CF_3COOH) are fully ionized in solution. Thus, the reaction shown above can be viewed as shown below, making it easier to identify the Lewis acid (s) and base(s).

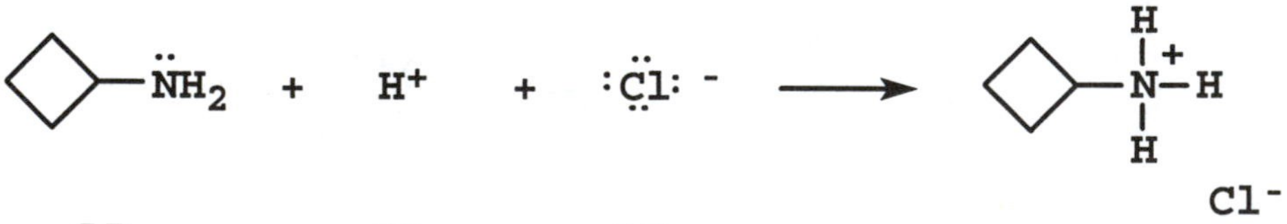

LB LA LB

(c) *Use curved arrows to indicate the flow of electrons from the atom that donates the pair of non-bonded or pi electrons to the receiving electron-deficient atom*, as illustrated below.

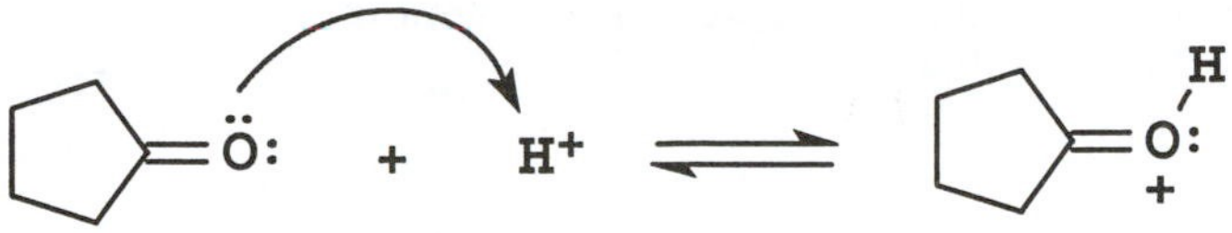

LB   LA

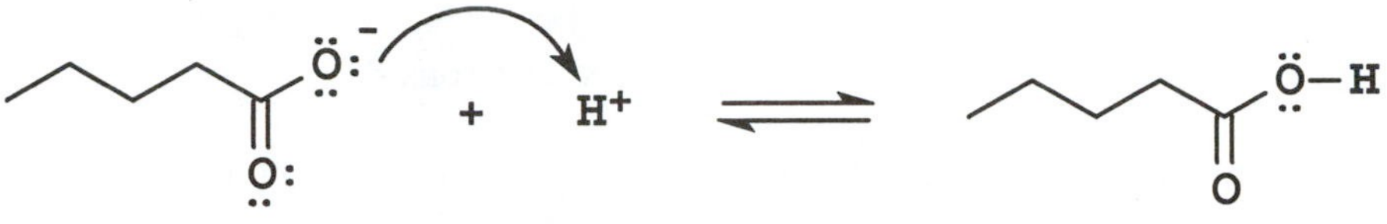

LB   LA

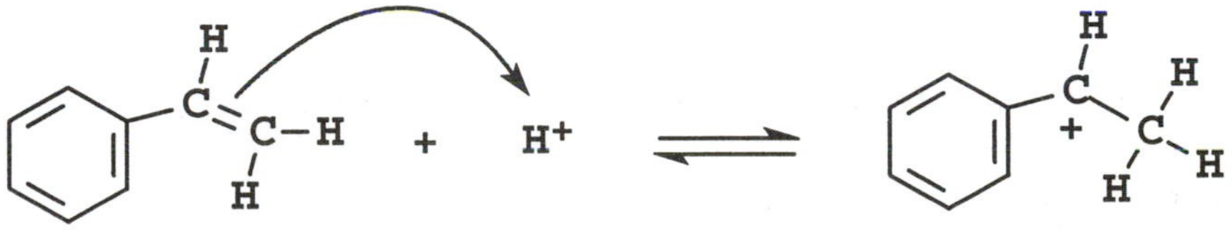

LB   LA

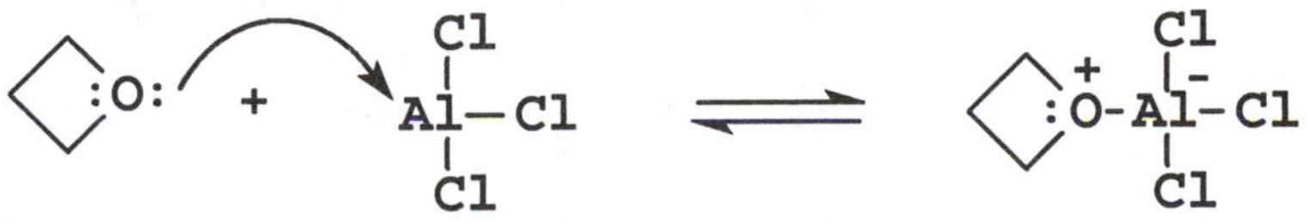

LB   LA

EXERCISE Use curved arrows to illustrate the flow of electrons in (b), examples 2-5.

(d) *Assign formal charges to the atoms directly involved in the reaction* (the formal charges on atoms that are not directly involved in the reaction do not change). Recall that the formal charge can be readily determined as follows:

$$\text{Formal charge on an atom} = X - Y - Z$$

$$\text{where } X = \text{number of valence electrons}$$

$$Y = \text{number of non-bonded electrons}$$

$$Z = \text{half the number of bonded electrons}$$

The number of valence electrons for an atom corresponds to its group number in the periodic table. For example, sulfur is in group VI in the periodic table, therefore it has six valence electrons (see also Minireview 1).

(e) If a reactant has more than one functional group which can donate a pair of non-bonded or pi electrons, then inspection of the structures of the reactant and product will usually reveal which functional group will react initially. Other considerations that come into play in deciding which functional group in a reactant is involved in the first step of a reaction include, for example, the stability of the *initial* carbocation. *The more stable a carbocation is, the easier it is to form* (see Minireview 3 for a discussion of carbocation chemistry). In the example shown below, while a pair of non-bonded electrons can be donated from either hydroxyl group, reaction takes place preferentially at the one that leads to the formation of the more stable benzylic carbocation, as opposed to the less stable primary carbocation.

benzylic carbocation

(f) Most mechanisms, particularly those involving *skeletal rearrangements*, may involve several sequential steps. Most of the steps are *consecutive Lewis acid/Lewis base reactions* which can be *intermolecular* (reaction involves two separate molecules that can be the same or different) or *intramolecular* (reaction takes place within the same molecule) in nature. The *driving force* behind these steps is the resulting *gain in stability* in going from one transient species to another such as, for example, from a less stable carbocation to a more stable carbocation, relief of ring strain, etc.

The following examples are meant to serve as a guide on how to write a reasonable mechanism for a reaction you may have never seen before using the Lewis acid/Lewis base approach. The same approach can be used to understand and explain how a reagent mediates a particular transformation. The thought processes underlying the general approach used in writing mechanisms are explicitly stated.

9

EXAMPLE 1 Write a mechanism for the following reaction:

$$\text{---}\!\!\!\!\!\!\overset{|}{\underset{|}{}}\text{---OH} \quad \xrightarrow{\text{HCl}} \quad \text{---}\!\!\!\!\!\!\overset{|}{\underset{|}{}}\text{---Cl} \quad + \quad \text{H}_2\text{O}$$

Answer

The first step in writing a mechanism for any chemical reaction is the *identification of the Lewis acid and Lewis base*. This can be readily accomplished by placing a sufficient number of non-bonded electron pairs on the heteroatom (oxygen) in the reactant to complete its octet (page 4). This immediately reveals that the alcohol will function as a Lewis base and the oxygen atom of the hydroxyl group will donate a pair of non-bonded electrons to a Lewis acid. Since mineral acids are fully-ionized in solution (page 6), the second reactant can be viewed as existing as H^+ (a Lewis acid) and Cl^- (a Lewis base) ions. Therefore, the first step of the mechanism for this reaction involves a Lewis base/Lewis acid reaction to yield a protonated alcohol. The second step involves the loss of a molecule of water via the cleavage of the C-O bond to yield a carbocation (a Lewis acid). Notice that the curved arrow is drawn in such a way so as to show that the pair of electrons that connects the two atoms in the C-O bond ends up on the oxygen atom of water as a pair of non-bonded electrons, at the same time relieving the positive charge on the electronegative atom. The second step in this mechanism is the rate-determining step, namely, the step with the highest free energy of activation). See Minireview 3 for a full discussion of carbocation chemistry). The third step completes the mechanism of the reaction. It involves a straightforward Lewis acid/Lewis base reaction between the

carbocation (LA) and the chloride ion (LB).

<u>EXAMPLE 2</u> Write a mechanism for the following reaction.

$$\text{(alkene)} \xrightarrow{\text{HBr}} \text{(alkyl bromide)}$$

<u>Answer</u>

By definition, any substance that can donate a pair of non-bonded or *pi electrons* is

a Lewis base (page 33). Thus, alkenes (as well as alkynes) invariably function as Lewis

bases by donating a pair of pi electrons to a Lewis acid. In the first step of this reaction,

the pair of pi electrons in the C=C bond is donated to H^+ (Lewis acid) to form a covalent

bond, yielding a carbocation (a Lewis acid). Notice that in this instance the pair of pi

electrons can potentially be used to form a covalent bond with either one of the carbon

atoms of the C=C bond. *In general, the reaction of an alkene with H^+ always leads to the*

initial formation of the most stable carbocation. In order to form the most stable

carbocation, H^+ will have to bond to the carbon of the C=C bond that bears the greater

number of hydrogens (Markovnikov's rule). The second step of the mechanism involves

the reaction of a carbocation (LA) with a Lewis base (bromide ion).

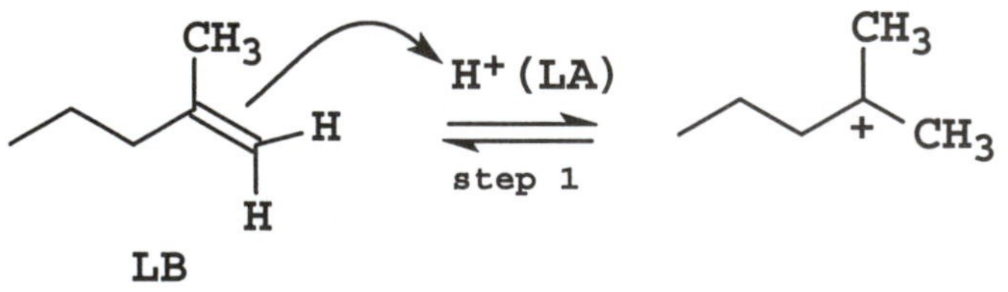

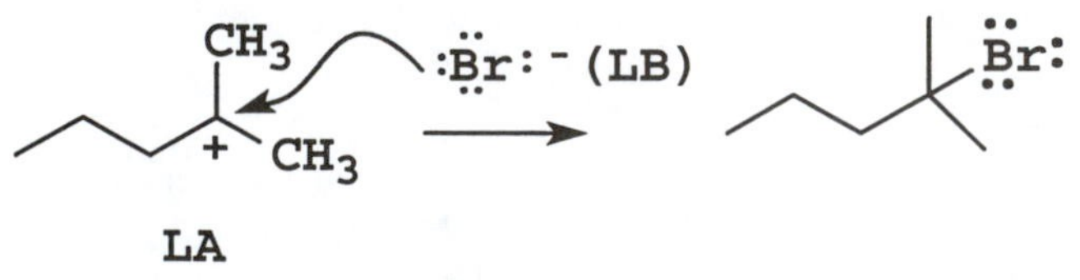

12

EXAMPLE 3 Write a mechanism for the following reaction.

Answer

Unlike Examples 1 and 2 which involved the reaction of either an alcohol or an

alkene with a Lewis acid, this example involves the participation of *two* Lewis bases (an

alkene *and* an alcohol) and a Lewis acid (H^+), leading to the formation of an ether. Prior

to writing a mechanism for this reaction, the following question must be addressed:

which one of the two bases will react with H^+ and how do you go about deciding that?

While the alcohol (LB) can potentially react with H^+ (LA), this is an unproductive

process (it cannot lead to the formation of the observed product). Note also that methyl

carbocations are highly unstable and do not form readily (see Minireview 3). Thus, the

reaction is initiated via a Lewis base/Lewis acid reaction between the alkene and H^+,

forming a secondary carbocation. Further reaction of the carbocation (a Lewis acid) with

the alcohol (a Lewis base) leads to the formation of a protonated ether (step 2). Loss of a

H^+ yields the observed product (step 3). The hydrogen in the protonated ether is *acidic*,

consequently step 3 is essentially an ionization step.

Exercise Write a mechanism for the following reaction.

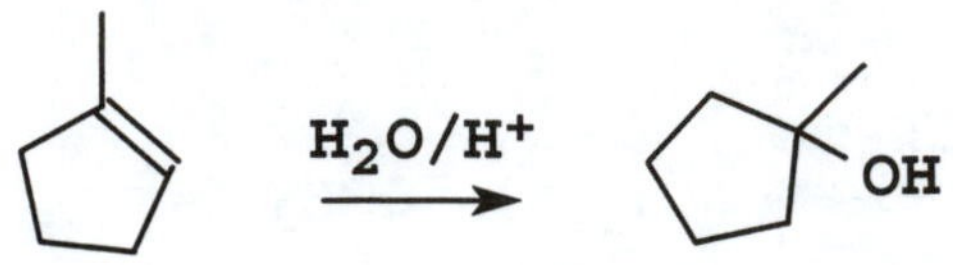

H$_2$O/H$^+$

EXAMPLE 4 Write a mechanism for the following reaction:

Answer

Once the appropriate number of pairs of non-bonded electrons is placed on the oxygen

to complete its octet, it can be readily seen that ethers (like alcohols), can function as

Lewis bases. Thus, the first step of this reaction is a Lewis acid/Lewis base reaction

leading to the formation of a protonated ether. This is followed by cleavage of the C-O

bond *in a way that yields the most stable carbocation.* A Lewis acid/Lewis base reaction

between the carbocation and the bromide ion completes the mechanism for this reaction.

<u>Note</u> Three and four-membered cyclic ethers readily undergo ring-opening reactions in

the presence of Lewis acids. The relief of ring strain serves as the driving force for these

reactions (see Minireview 6).

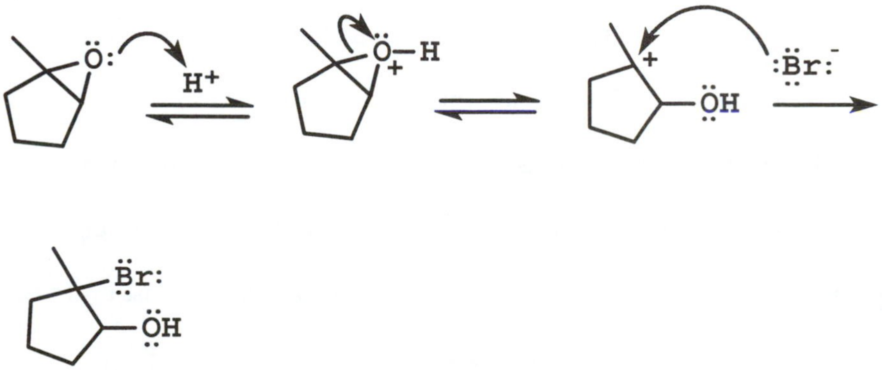

(mixture of
 diastereomers)

EXAMPLE 5 Write a mechanism for the following reaction:

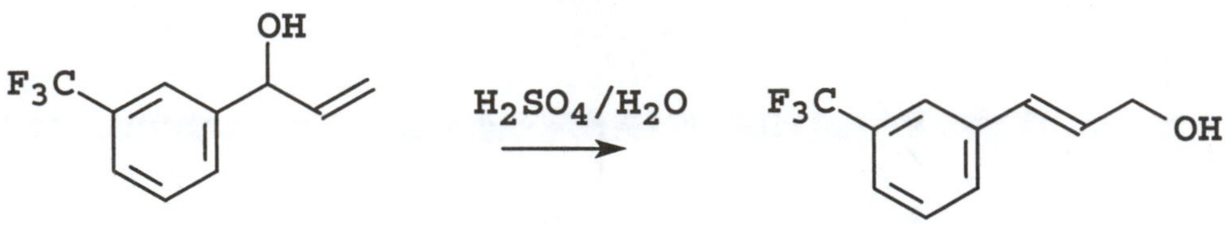

Answer

Once the Lewis acid and the Lewis base are identified, casual inspection of the Lewis base indicates that, in principle, three different functional groups (OH, alkene C=C bond and aromatic ring C=C bonds) can act as a Lewis base, namely, donate a pair of non-bonded or pi electrons to the Lewis acid (H^+). At this point a decision has to be made as to which of these will react with the Lewis acid. *As a general rule, it is useful to remember that (a) non-bonded electrons are more available for donation than pi electrons (since pi electrons are held by two nuclei) and, (b) an alkene C=C bond is more reactive than the C=C bonds of an aromatic ring*, since reaction of an aromatic ring C=C bond results in the loss of the aromatic character of the ring (~30 kcal/mol loss in resonance stabilization energy).

Thus, in Step 1 of this reaction the hydroxyl group of the Lewis base donates a pair of non-bonded electrons to the Lewis acid, which is then followed by cleavage of the C-O bond and the loss of a molecule of water (step 2), leading to the formation of a carbocation (notice that when the C-O bond is cleaved, the pair of electrons making up that bond goes with the oxygen). Step 1 (LB/LA reaction) and Step 2 (loss of water) are common to all reactions involving an alcohol and acid.

16

The carbocation formed in step 2 is resonance-stabilized, thus, reaction of the carbocation (LA) with water (LB) gives rise to a protonated alcohol (Step 3). The final step (Step 4) in this reaction involves the loss of a hydrogen ion (H^+) to give the observed product. Step 4 is a simple ionization step, analogous to the ionization of the hydronium ion ($H_3O^+ = H_2O + H^+$).

EXAMPLE 6 Write a mechanism for the reaction shown below.

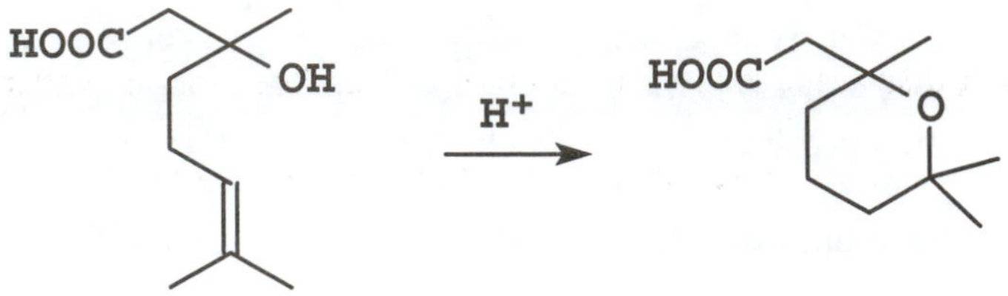

Answer

It's always a good idea to inspect the structures of the starting material and product, since in many instances this will quickly reveal which functional group(s) are involved in the reaction. In this instance, it is evident that it's the hydroxyl group and the C=C bond, and not the carboxyl group. With the non-bonded electrons added on the heteroatoms, and recalling the fundamental definition of a Lewis base, it is apparent that either the hydroxyl group or the C=C bond will donate a pair of electrons to the Lewis acid. In Example 1, it was stated that, as a general rule, a pair of non-bonded electrons is more available for donation than a pair of pi electrons. This example was chosen to (a) demonstrate that this is not always true, and (b) emphasize the need for you to keep an open (flexible) mind as you consider plausible mechanistic pathways. In other words, organic reactions frequently follow an unpredictable course, and the task on hand is to use fundamental principles to account for the formation of the observed product. Indeed, herein lies the pedagogical value of writing mechanisms. In so doing, you'll be forced to look at a situation in many ways and consider plausible pathways within a framework of principles. Thus, in Step 1 of this reaction, a Lewis base (C=C)/Lewis acid (H$^+$) reaction gives rise to a 3° carbocation (Markovnikov's rule). In Step 2, the carbocation (LA) accepts a pair of non-bonded electrons from the Lewis base (OH) to form a product that simply ionizes to give the observed product.

Notice that an initial reaction between the hydroxyl group and H^+, followed by loss of a molecule of water *also* leads to the formation of a 3° carbocation, however, this pathway can not account for the observed product (unproductive pathway).

EXAMPLE 7 Write a mechanism for the following reaction:

(2 mols)

Answer

As stated previously (Example 2), the C=C bond of an alkene functions as a Lewis

base by donating a pair of pi electrons to a Lewis acid. However, in this instance the pi

bonds in the phenyl (aromatic) ring could potentially behave the same way. Since the

reaction of a ring C=C with a Lewis acid would result in the formation of a much less

stable (non-aromatic) species, the alkene C=C bond reacts preferentially (step 1). The

carbocation formed is a Lewis acid which then reacts with a second molecule of the

alkene (a Lewis base) to give rise to a second benzylic carbocation (step 2). Besides

reacting with Lewis bases, a carbocation can lose a H^+ from an adjacent carbon atom to

form an alkene (see Minireview 3). Thus, step 3 leads to the formation of a new alkene.

20

EXAMPLE 8 Write a mechanism for the following reaction:

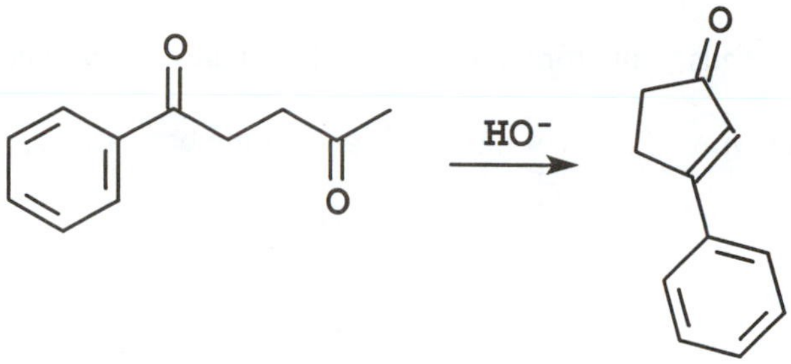

Answer

A mechanism question must always be approached from first principles, namely, it's not necessary for you to be able to realize that this particular reaction for example, is an aldol condensation reaction, in order to write a reasonable mechanism. Thus, the approach is always the same: first add non-bonded electron pairs to the two reactants, and classify each as a Lewis base and a Lewis acid. The stronger Lewis base (HO⁻, since it has a negative charge) is going to react with the second reactant which, by necessity, must function as a Lewis acid (LA). The carbon of the C=O group is electron-deficient (oxygen is more electronegative than carbon, consequently the electrons connecting the carbon and oxygen are not equally shared, and hence carbon has a partial positive charge (δ+) and oxygen a partial negative charge(δ-)). Thus, one possibility is for the hydroxide ion to donate a pair of electrons to the electron-deficient carbon (nucleophilic addition) or, since the hydrogens on the α carbon of a ketone are acidic, a Bronsted acid-base reaction can take place instead, yielding an anion (a Lewis base or nucleophile). *As a general rule, Bronsted acid-base reactions are faster than most other types of organic reactions*. Thus, a Bronsted acid-base reaction in Step 1 yields an anion, which then reacts in a Lewis base/Lewis acid reaction (Step 2) to form a product. This product is the conjugate base of an alcohol and can be viewed as being in equilibrium with the acid

(Step 3). Step 4 is a β elimination reaction, that leads to the formation of the product. Step 4 is facile because it leads to the formation of a highly stable *conjugated system* (a system that consists of an array of alternating double and single bonds).

EXAMPLE 9 Write a mechanism for the following reaction:

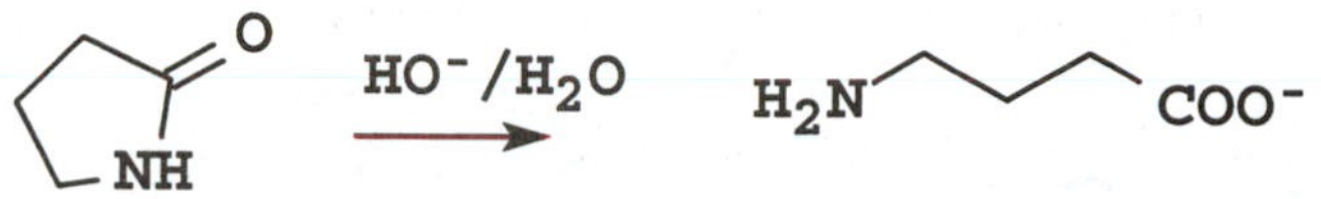

Answer

The carbonyl carbon is electron deficient, i.e., has a partial positive charge $(+\delta)$,

because it's bonded to two electronegative atoms. Any atom that bears either a full

positive charge (such as a carbocation, for example) or a partial positive charge, is

capable of accepting a pair of electrons from a Lewis base to form a covalent bond. Thus,

an initial LA/LB reaction leads to the formation of a tetrahedral intermediate (step 1).

Subsequent collapse of this intermediate leads to a ring-opened product (step 2). The

ring-opened product has an acidic group (COOH) and a basic group, thus a fast Bronsted

acid/base reaction (H^+ transfer) takes place, leading to the formation of the product (step

3).

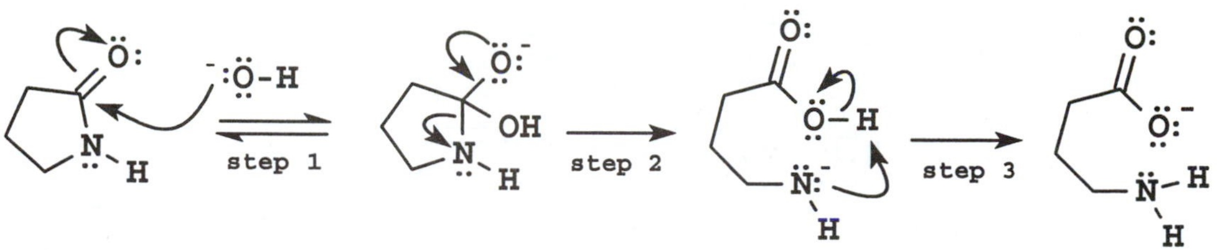

EXAMPLE 10 Write a mechanism for the following reaction:

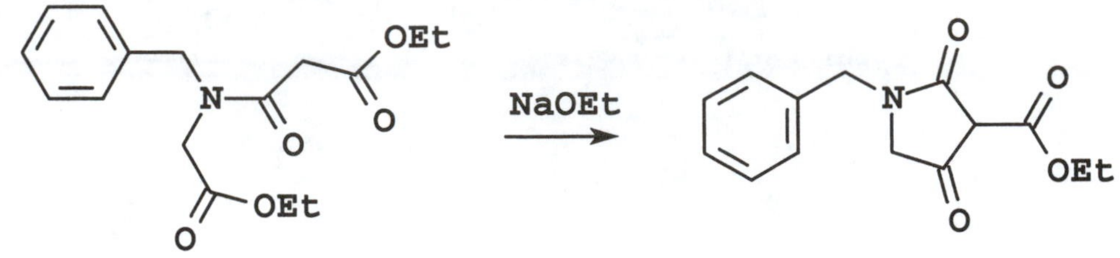

Answer

The Lewis base (ethoxide ion, $CH_3CH_2O^-$) reacts with the reactant to generate an anion (Step 1). Notice that the ethoxide ion reacts with the *most acidic hydrogen*. The pK_a of the reactant acid is ~11, while that of the product acid ($HOCH_2CH_3$) is ~16. Thus, the equilibrium lies to the right, i.e., favors the formation of the anion derived from the stronger acid (see Minireview 4 for details). Once the anion (nucleophile) is formed, an intramolecular LB/LA reaction (nucleophilic acyl substitution) takes place (Steps 2 and 3), forming the product.

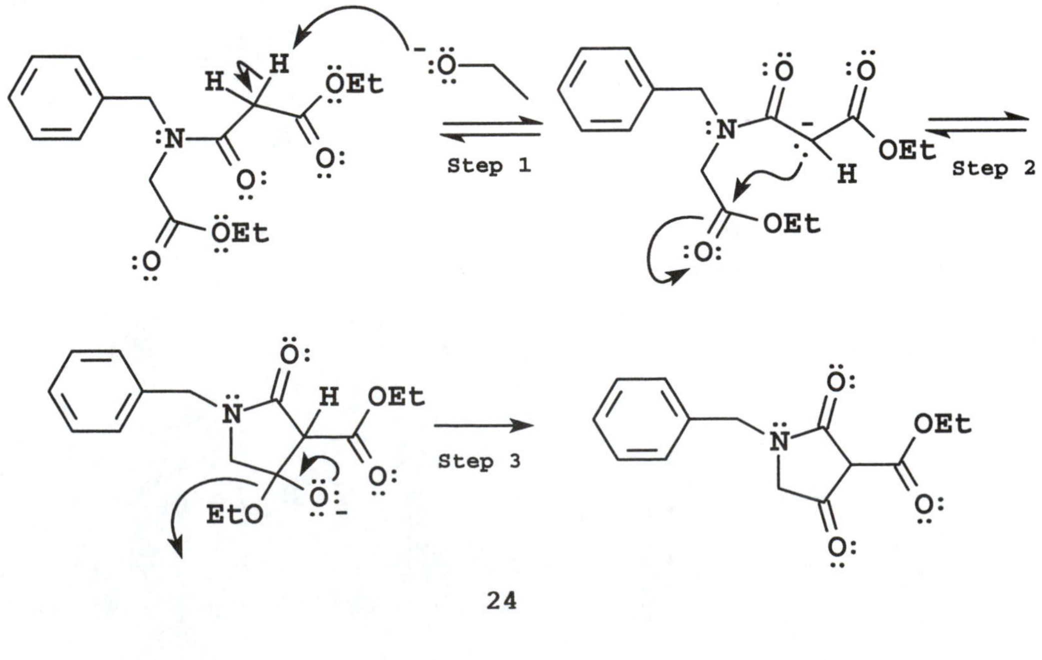

24

PART A

Minireviews 1-3 are intended to provide a quick review of the fundamental principles related to Lewis structures, Lewis acid/Lewis base reactions, resonance, and carbocation chemistry. These should be studied prior to attempting questions 1-34, Part A.

MINIREVIEW 1   Lewis Structures, Lewis Acids and Bases, and Lewis Acid/Lewis Base

Reactions

A sound understanding of mechanistic organic chemistry requires a proficiency in writing Lewis structures. Without the ability to draw Lewis structures correctly and with facility, a student is so severely handicapped that he or she will ultimately resort to learning organic chemistry by rote (a tedious, frustrating, and minimally-successful endeavor). The importance of this will become apparent momentarily.

A *Lewis structure* is a type of structural formula that shows the way in which atoms are bonded together, and depicts the bonding between atoms using pairs of non-bonded electrons (shown as dots) and bonded electrons (shown as dashes). In writing Lewis structures, the general approach outlined below should be followed:

1) *Determine the total number of valence electrons*. For neutral molecules, this is simply accomplished by adding up the valence electrons of the individual atoms. In the case of ions, an electron is added for each negative charge (anions), and an electron is subtracted for each positive charge (cations). *Recall that the number of valence electrons for an element corresponds to the group number of that element in the periodic table*. For example, nitrogen has five valence electrons (nitrogen is located in group five of the periodic table), fluorine has seven electrons (fluorine is located in group seven), etc.

2) *Connect the atoms in the given molecular formula using single lines* (dashes). It's helpful to remember that in the case of polyatomic molecules or ions, the atom of lower electronegativity is typically the central atom.

3) *Place a sufficient number of non-bonded electron pairs around each atom to give each atom an octet of electrons (octet rule).* Recall that hydrogen can only share a pair of electrons. If at this point the total number of valence electrons used is greater than that computed in step 1 above, use double or triple bonds to arrive at a Lewis structure that has the correct total number of valence electrons and all the atoms have an octet of electrons.

4) *Determine the formal charge on each atom.* As stated earlier, the formal charge can be readily determined as follows:

$$\text{Formal charge} = X - Y - Z$$

where X = number of valence electrons (of atom under consideration)

Y = number of non-bonded electrons, and

Z = half the number of bonded electrons

NOTES

a) The Lewis structures of compounds derived from group IIIA elements (B, Al, etc.) have incomplete octets. As might be expected, these compounds invariably function as Lewis acids and are also highly reactive (because of their incomplete octet, they tend to readily accept a pair of electrons from a Lewis base, thereby acquiring an octet of electrons).

b) Many chemical reactions proceed through the transient formation of highly reactive species. For example, carbocations, as well as other species that lack an octet of electrons, have a high energy (low stability) and, consequently, are highly reactive.

c) Lewis structures in which all the atoms have an octet of electrons cannot be written for molecules and ions that have an odd total number of valence electrons (for example, nitric oxide (NO). As might be expected, such species also exhibit high chemical reactivity.

d) The atoms of elements that have empty d orbitals can expand their octets, namely, they can accommodate more than eight electrons. Sulfur and phosphorus are the two elements most commonly encountered in organic chemistry.

In summary, species that lack an octet of electrons (electron-deficient species) have two distinct characteristics: they function as Lewis acids and are highly reactive (vide infra).

EXAMPLES

CCl_4 1   C   1 x 4 = 4
              4   Cl 4 x 7 = 28
                            ___
                            32
                  (total no. of valence electrons)

$$:\overset{..}{\underset{..}{Cl}}:$$
$$:\overset{..}{\underset{..}{Cl}}-\overset{|}{\underset{|}{C}}-\overset{..}{\underset{..}{Cl}}:$$
$$:\overset{..}{\underset{..}{Cl}}:$$

PH_3 P   1 x 5 = 5
           H   3 x 1 = 3
                         _
                         8

$$H-\overset{\overset{H}{|}}{\underset{\underset{H}{|}}{P}}:$$

C_2H_4 C   2 x 4 = 8
           H   4 x 1 = 4
                         __
                         12

$$\overset{H}{\underset{H}{}}\!\diagdown C=C\diagup\overset{H}{\underset{H}{}}$$

29

$SOCl_2$

S 1 x 6 = 6
O 1 x 6 = 6
Cl 2 x 7 = 14

$\overline{26}$

$$:\!\overset{..}{\underset{..}{Cl}}\!-\!\overset{\overset{\displaystyle \overset{..}{O}:}{\|}}{\underset{}{S}}\!-\!\overset{..}{\underset{..}{Cl}}\!:\quad \text{or} \quad :\!\overset{..}{\underset{..}{Cl}}\!-\!\overset{\overset{\displaystyle :\overset{..}{O}:^-}{|}}{\underset{}{\overset{+}{S}}}\!-\!\overset{..}{\underset{..}{Cl}}\!:$$

N_2

N 2 x 5 = 10

$$:N\equiv N:$$

BF_3

B 1 x 3 = 3
F 3 x 7 = 21

$\overline{24}$

$$:\overset{..}{\underset{..}{F}}\!-\!\overset{\overset{\displaystyle :\overset{..}{F}:}{|}}{\underset{\underset{\displaystyle :\overset{..}{F}:}{|}}{B}}$$

NOCl

N 1 x 5 = 5
O 1 x 6 = 6
Cl 1 x 7 = 7

$\overline{18}$

$$:\overset{..}{\underset{..}{Cl}}\!-\!\overset{..}{N}\!=\!\overset{..}{O}:$$

CH_3NO_2

C 1 x 4 = 4
H 3 x 1 = 3
N 1 x 5 = 5
O 2 x 6 = 12

$\overline{24}$

$$\overset{\displaystyle H}{\underset{\displaystyle H}{H\!-\!C}}\!-\!\overset{+}{N}\!\!\overset{\overset{\displaystyle \overset{..}{O}:}{\diagup}}{\underset{\underset{\displaystyle :\overset{..}{O}:^-}{\diagdown}}{}}\quad \text{or} \quad \overset{\displaystyle H}{\underset{\displaystyle H}{H\!-\!C}}\!-\!\overset{+}{N}\!\!\overset{\overset{\displaystyle :\overset{..}{O}:^-}{\diagup}}{\underset{\underset{\displaystyle \overset{..}{O}:}{\diagdown}}{}}$$

CH_2O C 1 x 4 = 4
     H 2 x 1 = 2
     O 1 x 6 = 6
          ————
          12

$$H-C\overset{\cdot\cdot}{\underset{\displaystyle H}{\Vert}}\overset{\cdot\cdot}{O}\colon$$

NO_2^- N 1 x 5 = 5
      O 2 x 6 = 12
            1 (one negative charge)
            ————
            18

$$^-\colon\!\overset{\cdot\cdot}{\underset{\cdot\cdot}{O}}-\overset{\cdot\cdot}{N}=\overset{\cdot\cdot}{O}\colon$$

CN^- C 1 x 4 = 4
     N 1 x 5 = 5
           1
           ——
           10

$$^-\colon\! C\equiv N\colon$$

CO_3^{2-} C 1 x 4 = 4
       O 3 x 6 = 18
             2 (two -ve charges)
             ————
             24

$$^-\colon\!\overset{\cdot\cdot}{\underset{\cdot\cdot}{O}}-\overset{\textstyle\overset{\cdot\cdot}{O}\colon}{\underset{\displaystyle}{C}}-\overset{\cdot\cdot}{\underset{\cdot\cdot}{O}}\colon^-$$

CH_3O^- C 1 x 4 = 4
       H 3 x 1 = 3
       O 1 x 6 = 6
             1 (one -ve charge)
             ————
             14

$$H-\overset{\textstyle H}{\underset{\textstyle H}{C}}-\overset{\cdot\cdot}{\underset{\cdot\cdot}{O}}\colon^-$$

$NaBH_4$ This is the same as $Na^+ BH_4^-$. Thus, the Lewis structure that we want is that of BH_4^-.

```
B   1 x 3 = 3                        H
H   4 x 1 = 4                        |  -
            1 (one -ve charge)   H—B—H
           ___                      |
            8                       H
```

NO^+
```
N   1 x 5 = 5                                    formal charge on O
O   1 x 6 = 6          :N≡O:            (6-2-3 = +1)
         ___
          11      - 1 (subtract one electron for each
         ___              positive charge)
          10
```

(There is an arrow pointing to the $+$ sign above O, with the label "formal charge on O (6-2-3 = +1)")

$C_2H_5^+$
```
C   2 x 4 = 8              H  H
  H   5 x 1 = 5           |  |
                      H—C—C+
           ___           |  |
            13           H  H
         - 1
           ___
            12
```

<u>Exercise</u> Write three Lewis structures for $C_3H_3^+$.

Lewis Acids and Bases, and Lewis Acid/Lewis Base Reactions

A Lewis base is substance that can donate a pair of non-bonded or pi electrons to a Lewis acid to form a covalent bond. Lewis bases that have a negative charge are *stronger* Lewis bases than those without a negative charge. For example, CH_3O^- is a stronger Lewis base than CH_3OH and, consequently, will react faster with a Lewis acid. Furthermore, Lewis basicity is directly related to the availability of the pair of non-bonded electrons. A non-bonded electron pair on a less electronegative atom is more available for donation than a non-bonded electron pair on a more electronegative atom. This is the reason why, for example, amines are stronger Lewis bases than alcohols and ethers. Likewise, aliphatic amines are stronger Lewis bases than aromatic amines (in aromatic amines the non-bonded electron pair on the nitrogen atom is delocalized over the aromatic ring via resonance and, consequently, is not as available for donation).

A Lewis acid is a substance that can accept a pair of non-bonded or pi electrons from a Lewis base to form a covalent bond. Lewis acids are electron-deficient, namely, an atom in a Lewis acid has a *positive charge*, or it may have a *partial positive charge* because it's bonded to one or more electronegative atoms. An atom with an incomplete octet of electrons also acts as a Lewis acid, readily accepting a pair of electrons from a Lewis base.

A Lewis acid/Lewis base reaction can be generally described as shown below.

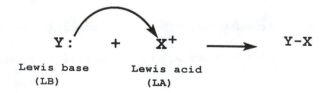

Lewis base (LB)   Lewis acid (LA)

When organic reactions are described this way, the product of an organic reaction can be readily predicted, *without recourse to memorization*. The following examples illustrate these ideas.

Example 1

$$HO^- \quad + \quad H^+ \quad \longrightarrow \quad H_2O$$

$$\text{or} \quad H-\ddot{O}\colon^- \quad + \quad H^+ \quad \longrightarrow \quad H-\ddot{O}-H$$

$$\qquad\qquad \text{LB} \qquad\qquad \text{LA}$$

Example 2

$$BH_3 \quad + \quad NH_3 \quad \longrightarrow \quad H_3\overset{-}{B}-\overset{+}{N}H_3$$

$$\text{or} \qquad\qquad H-B \quad + \quad \colon N-H \quad \longrightarrow \quad H-B-N-H$$

$$\qquad\qquad\qquad \text{LA} \qquad\qquad \text{LB}$$

Notice that the overall process involves

(a) *writing the Lewis structures of the two reactants and identifying the reactant which functions as a Lewis acid and the reactant that functions as a Lewis base;*

(b) *using a curved arrow to show how the non-bonded or pi electron pair of the Lewis base is used to form a covalent bond with the Lewis acid and,*

(c) *determining the formal charges on the two atoms involved in the formation of the new covalent bond.*

Some additional examples illustrating this process are given below:

Example 3

$$CH_3OH \quad + \quad HBr \longrightarrow CH_3OH_2^+ \quad + \quad Br^-$$

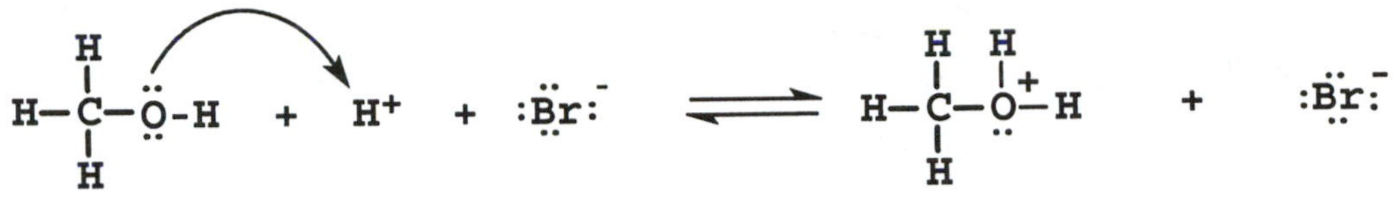

NOTES (1) As mentioned earlier, because strong mineral acids (HCl, HBr, HNO₃, etc.) and strong organic acids (RSO₃H, RCOOH, etc.) are *ionized* in solution, it's best to represent them as ions. When written that way, it can be readily ascertained which is the Lewis acid and which is the Lewis base. Strong bases and ionic compounds should be treated the same way. (2) Recall that a Lewis base will only react with a Lewis acid, but *not* with another Lewis base.

Example 4

$$CH_3NH_2 \quad + \quad HCl \longrightarrow CH_3NH_3^+ \quad + \quad Cl^-$$

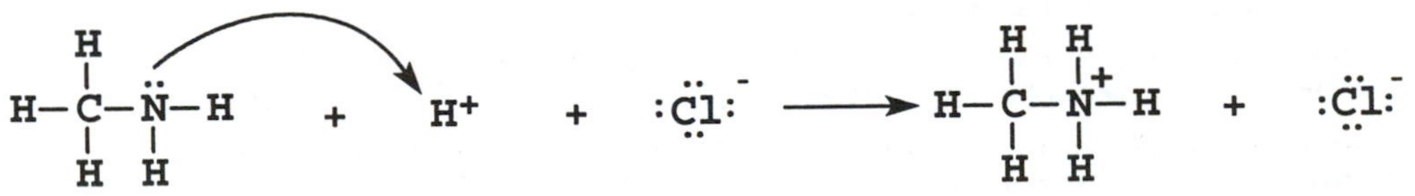

35

Example 5

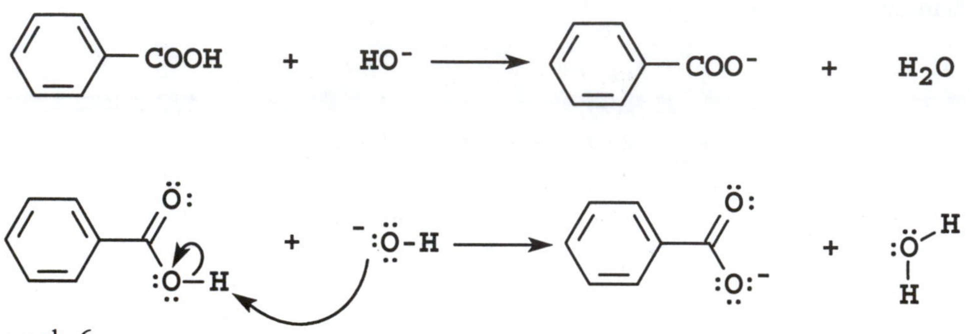

Example 6

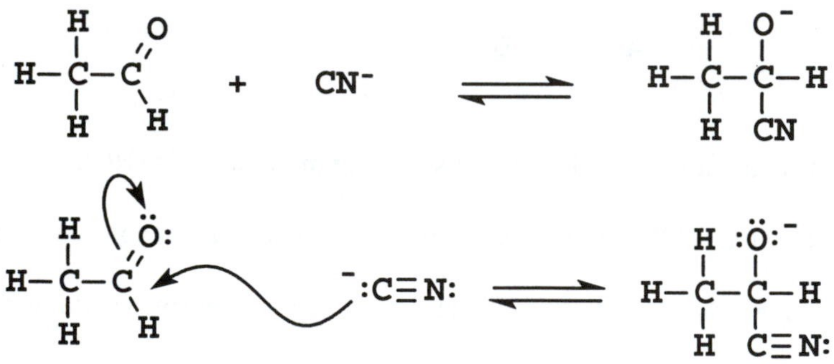

LA   LB

Notice that in this example the carbonyl carbon is electron-deficient because it's

bonded to an electronegative atom and, consequently, has a partial positive charge ($\delta+$).

Thus, the molecule behaves as a Lewis acid and accepts a pair of electrons from the

Lewis base. In general, an atom that is bonded to one or more electronegative atoms will

have a partial positive charge and can, in principle, function as a Lewis acid by accepting

a pair of electrons from a Lewis base.

In the familiar S_N2 reaction, the Lewis base (or *nucleophile*) donates a pair of

electrons to the electron-deficient atom, namely, the carbon bonded to the electronegative

atom, with simultaneous cleavage of the carbon-halogen bond (illustrated in example 7).

36

<u>Example 7</u>

$$HO^- \quad + \quad CH_3I \quad \longrightarrow \quad CH_3OH \quad + \quad I^-$$

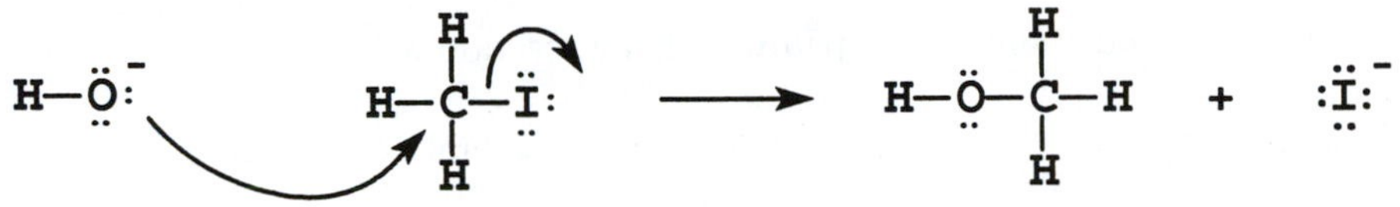

LB   LA

MINIREVIEW 2   Resonance

A general familiarity with the concept of resonance and a facility in writing resonance structures is very helpful in (a) assessing the stability of an individual species (anion, cation, or radical), (b) assessing the relative stability of two similar species, thereby enabling one to predict the pathway that a reaction will likely to follow and/or which species is likely to form (since the greater the stability of a species is, the easier it is to form) and, (c) predicting the site of a reaction in a molecule.

When a molecule or ion can be represented by two or more Lewis structures that differ only in the positions of the electrons, (a) none of those Lewis structures represents the *actual* structure of the molecule or ion and, (b) the actual structure of the molecule or ion is best represented by a hybrid (called *resonance hybrid*) of these resonance structures. *Resonance involves the alternate placement of non-bonded or pi electrons over the same atomic skeleton, without any change in the positions of the atoms.*

In writing resonance structures, the following general rules should be followed:

(1) *the greater the number of resonance structures that can be written for a species, the more stable that species will be;*

(2) *resonance structures in which all the atoms have an octet of electrons are more stable;*

(3) *other things being equal, a resonance structure with a negative charge on the most electronegative atom will have greater stability. Conversely, a resonance structure with a positive charge on the least electronegative atom will be more stable.*

(4) *Maximum stabilization of a species (anion or cation) is achieved when the contributing resonance structures are equivalent, namely, they have the same energy.*

Examples illustrating these rules are given below.

<u>Resonance Structures Involving Anions</u>

<u>Example 1</u> The two resonance structures that can be written for the carboxylate anion

(RCOO⁻) are shown below.

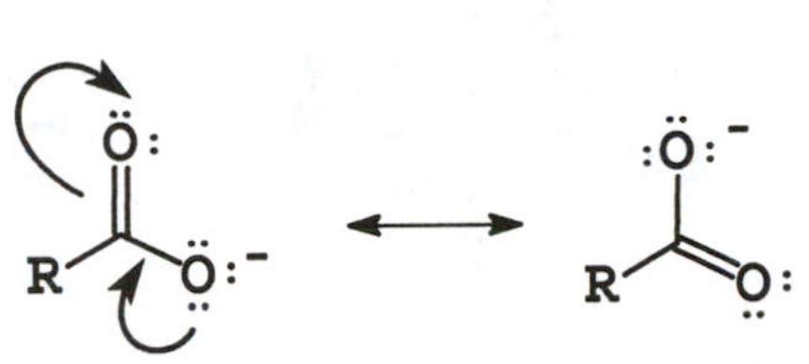

Note that these structures are equivalent, namely, they have the same energy (they both

have a negative charge on the same kind of atom) [Rule 4]. Notice also that the

delocalization of the negative charge over the two oxygen atoms via resonance stabilizes

the anion and is the reason for the observed acidity of carboxylic acids. In other words,

the hydrogen in a carboxylic acid (RCO<u>OH</u>) is acidic and readily donated to a base since

a resonance-stabilized anion is formed in the process (see also Minireview 4 for a full

discussion of the relationship between Bronsted acid strength and resonance stabilization

of the conjugate base).

<u>Exercise</u> Sulfonic acids (RSO₃H) are stronger acids than carboxylic acids. Use your

knowledge of resonance to account for this observation.

<u>Example 2</u>

The resonance structures shown above are *non-equivalent*, since the negative charge

is on two different kinds of atoms. The resonance structure with the negative charge on the more electronegative atom is more stable [Rule 3]. Recall that electronegativity follows the order $F > O > N > Cl > C$.

Example 3

(I)   (II)   (III)

The order of stability in resonance structures A-C parallels the order of electronegativity of the three atoms bearing the negative charge [Rule 3]. Thus,

$$(II) > (III) > (I)$$

$$\longleftarrow$$

increasing stability

$$O > N > C$$

$$\longleftarrow$$

increasing electronegativity

As stated earlier, the greater the number of resonance structures that can be written for an anion, the greater the stability of the anion [Rule 1]. Put differently, *the greater the delocalization of the negative charge, the greater the stability of an anion.* Consequently, in comparing the relative stability of two anions, the more stable anion will be the one for which a greater number of resonance structures can be written. Corollary: resonance-stabilized anions have higher stability and, therefore, are easier to form. Thus, the relative acidity of a Bronsted acid is determined by the stability of the corresponding anion.

40

Example 4 When ethyl acetoacetate is treated with base it readily forms the resonance-stabilized anion shown below:

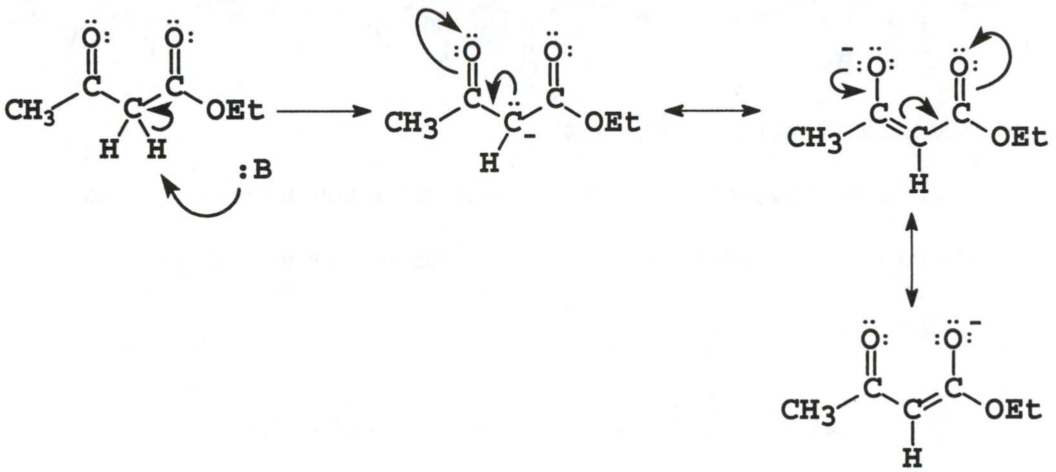

Exercise (a) Rank the resonance structures in example 4 in order of decreasing stability (most stable first). (b) The three methyl hydrogens in ethyl acetoacetate are *less* acidic than the two methylene hydrogens. Why?

Example 5 Assess the relative stability of anions A and B below:

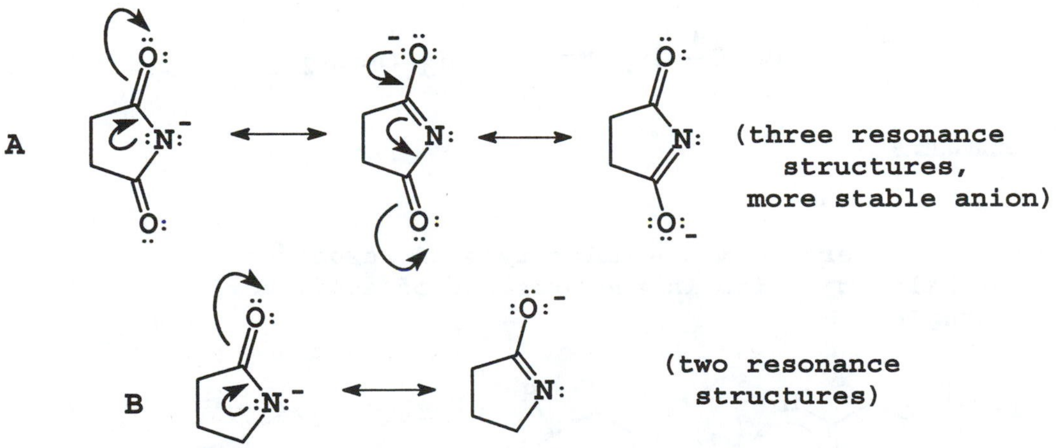

A   (three resonance structures, more stable anion)

B   (two resonance structures)

Three resonance structures can be written for anion A versus two resonance structures for anion B. Hence, anion A is more stable than anion B [Rule 1].

Example 6 Write all the major resonance structures for the anion shown below.

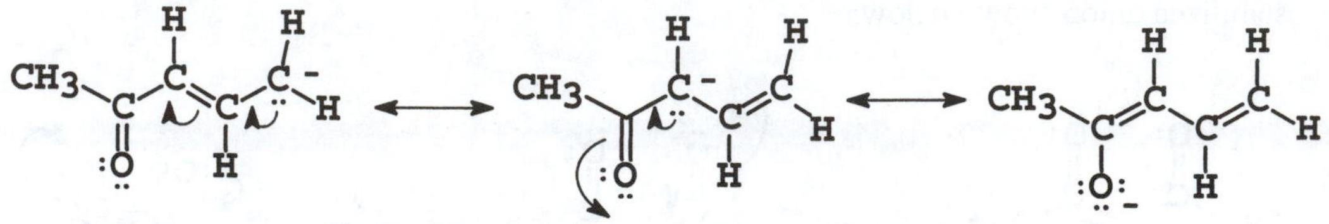

Resonance Structures Involving Cations

Allylic and benzylic carbocations, as well as carbocations having a heteroatom directly bonded to the carbon bearing the positive charge, are stabilized by resonance.

Example 1

$$CH_3-CH=CH-\overset{+}{C}H_2 \longleftrightarrow CH_3-\overset{+}{C}H-CH=CH_2$$

Example 2

Example 3

$$CH_3-\ddot{\overset{..}{O}}-\overset{+}{C}H_2 \longleftrightarrow CH_3-\overset{+}{O}=CH_2$$

Example 4

Vinyl ethers form a similar type of resonance-stabilized cation in the presence of acid. For example,

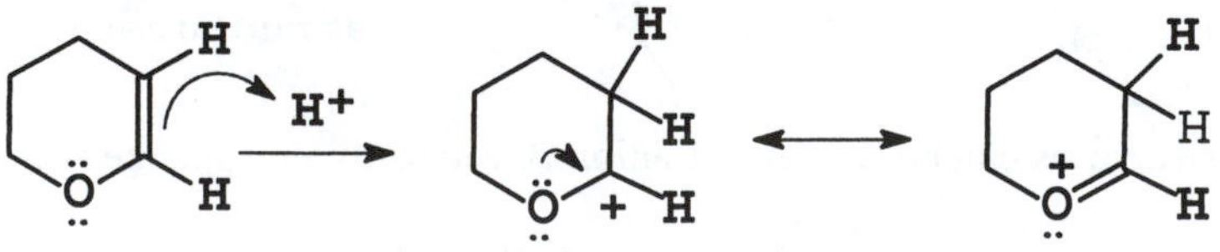

42

Example 5

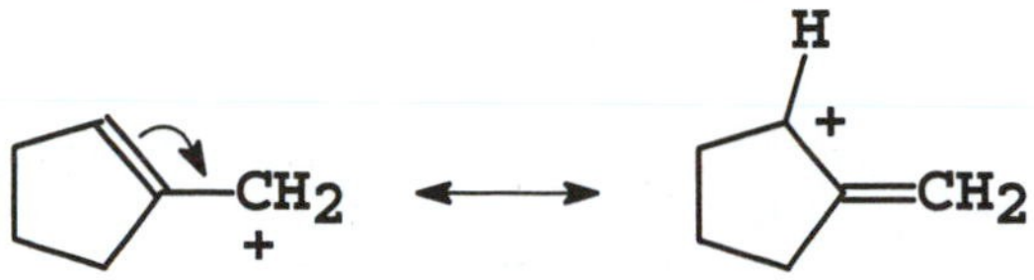

Example 6

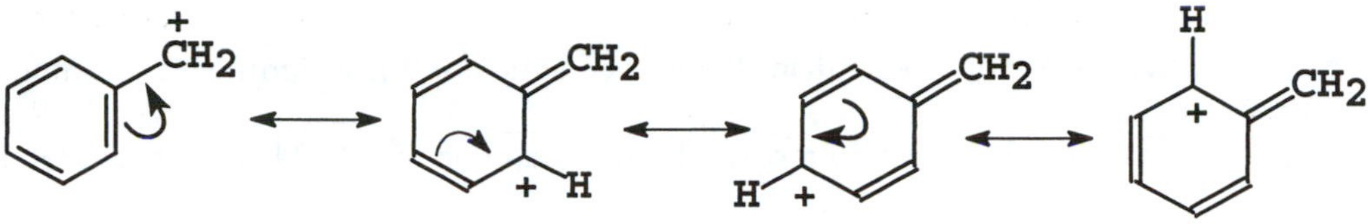

Example 7

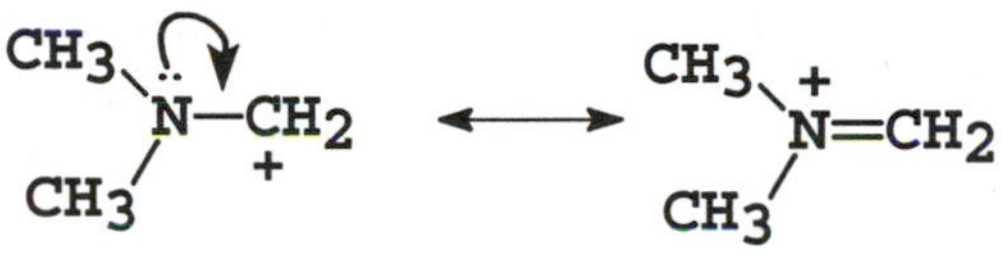

Example 8

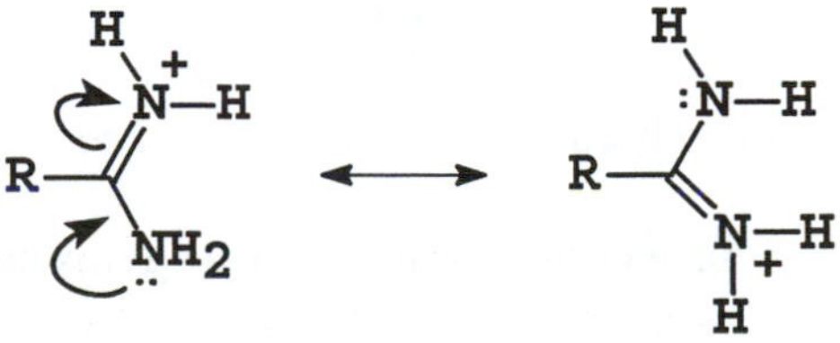

43

Resonance Structures Involving Neutral Molecules

Example 1

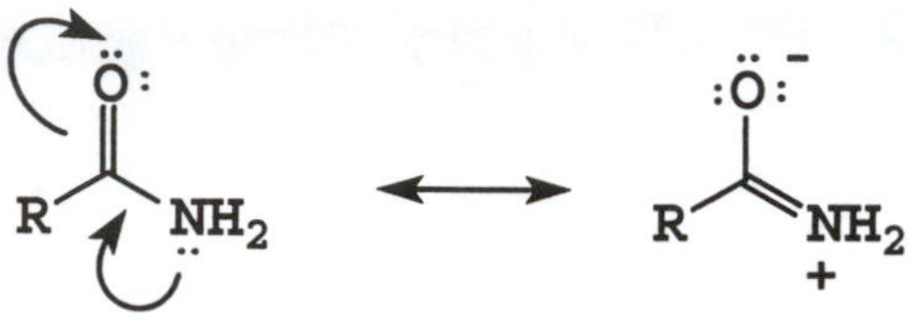

A   B

Note Structure A is more stable than B, however, structure B is a significant contributor that accounts for (a) the observed restricted rotation around the C – N bond in amides and peptides and, (b) the decreased Lewis basicity of amides versus amines (arising from the lower availability of the non-bonded electron pair on the nitrogen in amides). In thiourea and thioamides, S_N2 reactions occur at sulfur because of the contribution of the resonance structures shown below, and the high nucleophilicity of sulfur (see Minireview 5 for a discussion of nucleophilicity and its relationship to basicity).

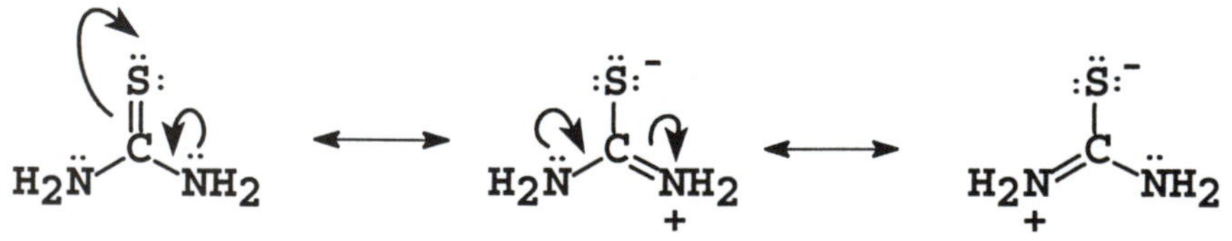

Resonance structures can be used to predict the *site of reaction* in neutral molecules. For example, the course of the reaction of an α,β-unsaturated compound with a Lewis base (Michael addition reaction) can be predicted by considering the resonance structure shown below. This resonance structure clearly indicates that the β carbon is electron-deficient (Lewis acid) and can accept a pair of electrons from a nucleophile (Lewis base).

44

Likewise, the following resonance structures identify the sites of reaction of vinyl ethers and enamines with Lewis acids and provide a better understanding of the chemical behavior of these classes of compounds.

Carbocation Chemistry

The reaction of an alkene, alkyne or alcohol (all Lewis bases) with H^+ (a strong

Lewis acid) leads to the initial formation of a carbocation. *Carbocations are transient,*

electron-deficient and highly-reactive species. Once formed, they function as Lewis acids

that react rapidly with Lewis bases (in the process the carbon atom bearing the positive

charge completes its octet). Some noteworthy characteristics of carbocations are

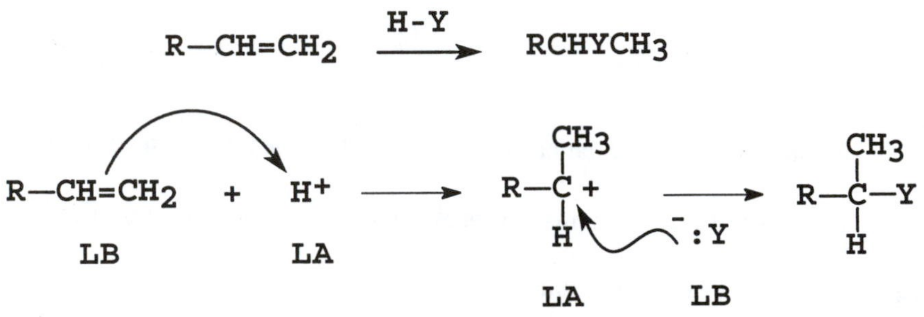

1) Carbocations vary in stability depending on their structure. *The greater the stability of*

a carbocation is, the easier it is to form. Thus, the initial carbocation formed in a given

reaction is invariably the one of highest stability. The *order of stability* in carbocations is

as follows:

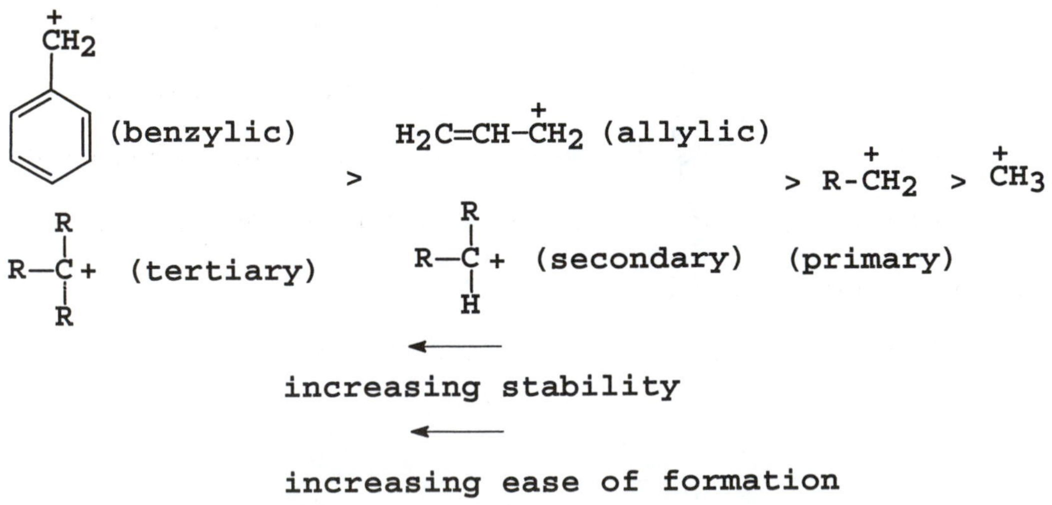

46

Vinyl and phenyl carbocations are highly unstable. Consequently, although the hydroxyl group of a phenol can, in principle, function as a Lewis base (just like the hydroxyl group of an alcohol), it never yields a phenyl carbocation.

As mentioned earlier, carbocations that have a positive charge on a carbon atom which is bonded to a heteroatom (O, N, S) are stabilized by resonance.

$$-\overset{+}{C}-\overset{..}{X}-R \longleftrightarrow -C=\overset{+}{X}-R \quad (X = O, S, N)$$

2) *Carbocations frequently undergo rearrangements via 1,2-hydride or 1,2-alkyl shifts to form carbocations of equal or greater stability.* These carbocations may on occasion arise via consecutive 1,2-hydride and/or 1,2-alkyl shifts. Further reaction with a Lewis base, or loss of a H^+ from an adjacent carbon (E_1), ultimately yields the observed product(s).

In certain cases, initial formation of a carbocation is followed by *ring expansion* to form a new carbocation. The *driving force* behind the observed ring expansion is the relief of ring strain and the formation of a more stable species. *Recall that as ring size decreases, ring strain increases.* Thus, 3- and 4-membered rings have considerable ring strain and tend to undergo ring expansion. *Recall also that as ring strain increases, chemical reactivity increases.*

An interesting variation of this theme involves the cyclopropylmethyl carbocation.

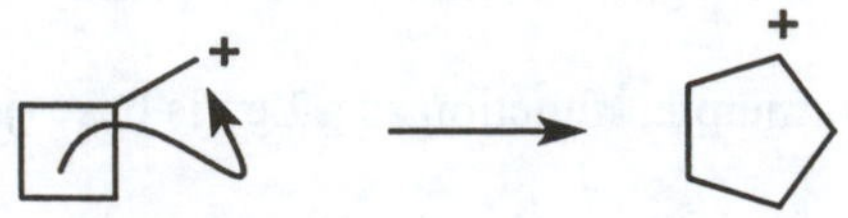

(relief of ring strain via ring expansion)

(homoallylic carbocation)

or

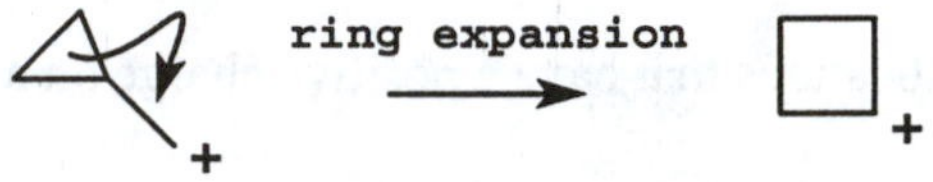

ring expansion

3) *Carbocations are planar (flat), since the carbon bearing the positive charge is sp^2 hybridized.* Consequently, *a carbocation derived from an optically active reactant will ordinarily yield an optically inactive product (a racemic mixture) upon reaction with a Lewis base.* Attack on an sp^2 hybridized carbon by a Lewis base is equally likely from either side, leading to a 1:1 mixture of the R and S isomers.

<u>Example</u> When (S)-2-butanol is treated with a trace amount of acid, it undergoes racemization. Write a plausible <u>mechanism</u> that accounts for this observation.

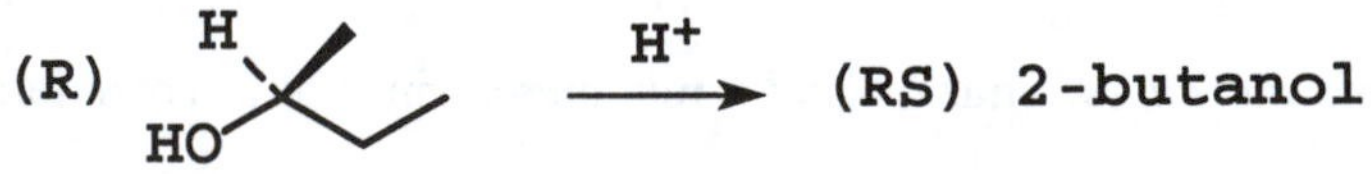

(R) ⟶ H^+ ⟶ (RS) 2-butanol

48

Formation of the planar carbocation is followed by a LB/LA reaction. Since attack by the Lewis base (H_2O) is equally likely from either side, the reaction generates a 1:1 mixture of the two enantiomers, namely, a racemic mixture.

4) An *electrophilic aromatic substitution* reaction can be viewed as a two-step Lewis base/Lewis acid reaction involving an aromatic compound (acting as a Lewis base) and a transiently-generated Lewis acid (also called an *electrophile*). Recall that the first step in an electrophilic aromatic substitution reaction is the *rate-determining step* (has the highest free energy of activation). Typical transient Lewis acids include carbocations and other electron-deficient species such as NO_2^+, Br^+, Cl^+, etc.

In the familiar *Friedel-Craft alkylation reaction* a carbocation is generated by mixing an alkyl halide (RX) with a Lewis acid (AlX_3), or by mixing an alcohol (ROH) or alkene with acid. As expected, the formation of a carbocation in this reaction frequently leads to the formation of rearranged products.

50

The reaction of an aromatic compound with an acid chloride in the presence of a Lewis acid is referred to as the *Friedel-Craft acylation reaction*. An initial Lewis acid/Lewis base reaction leads to the transient formation of a resonance-stabilized carbocation (also called an *acylium ion*) which then reacts with the aromatic compound (illustrated below). *In contrast to ordinary carbocations, acylium ions do not undergo rearrangements.*

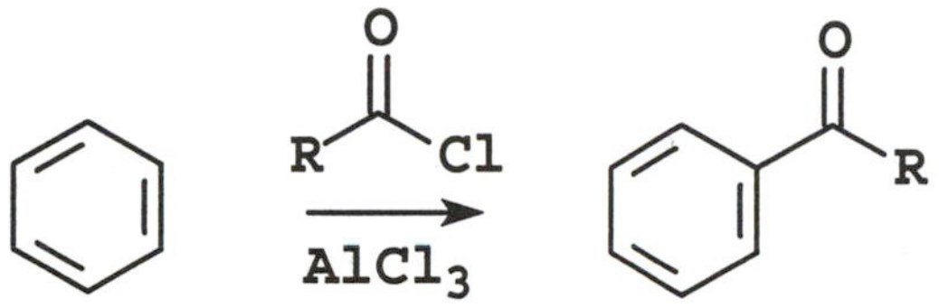

Mechanism

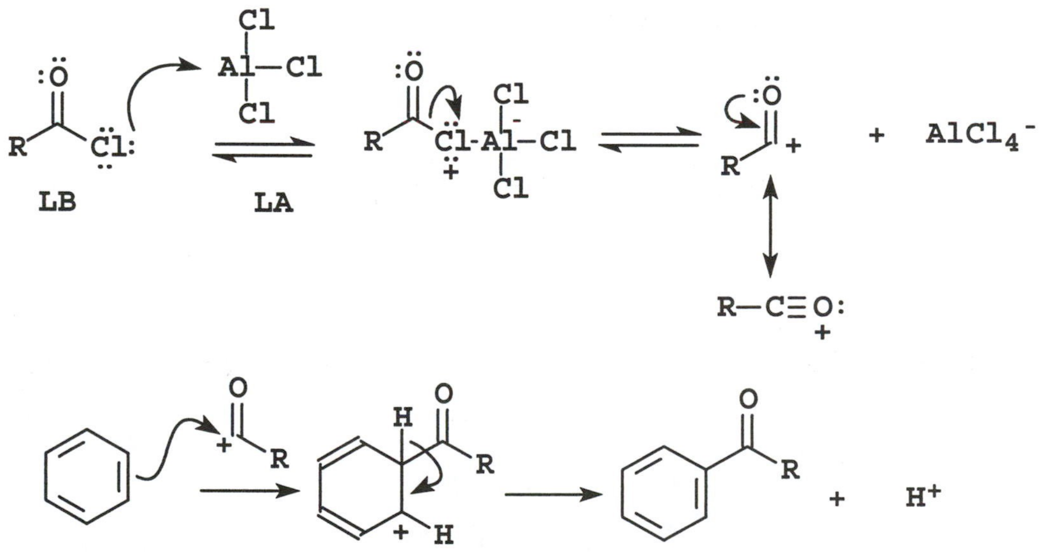

In the nitration and halogenation reactions the Lewis acid (electrophile) is transiently generated via a sequence of Lewis acid/Lewis base reactions using concentrated HNO_3/H_2SO_4 and Br_2 (or Cl_2) with $FeBr_3$ (or $FeCl_3$), respectively. Recall that the *presence and nature of a substituent* on the aromatic ring have a profound effect on *reactivity and orientation. Electron-donating groups* (R, OR, NHCOR, OH, NH$_2$, etc.) enhance the Lewis basicity and, hence, the reactivity of an aromatic compound. Furthermore, these groups direct the Lewis acid (electrophile) to the *ortho* and *para* positions. In contrast, *electron-withdrawing groups* (NO$_2$, CN, CHO, COR, COOR, etc.) decrease reactivity and direct the electrophile to the *meta* position. Halogens are deactivating, but ortho and para-directing.

QUESTIONS 1-34

Questions 1-34 aim at reviewing and gaining a better understanding of the following topics:

*Lewis structures

*Lewis acids and bases

*Lewis acid/Lewis base reactions

*Resonance (rules of resonance, writing resonance structures, assessment of relative stability of anions and cations using resonance).

*Carbocation Chemistry (formation, stability, rearrangements, stereochemistry and reactions of carbocations)

*Lewis acid/Lewis base reactions of alcohols, alkenes, alkynes and epoxides.

*Lewis acid/base reactions involving aromatic rings (electrophilic aromatic substitution reactions, including Friedel-Craft alkylations and acylation reactions).

1)

2)

3)

4)

5)

6)

7)

8)

9)

10)

11)

12)

56

13)

14)

15)

16)

17)

18)

19)

20)

21)

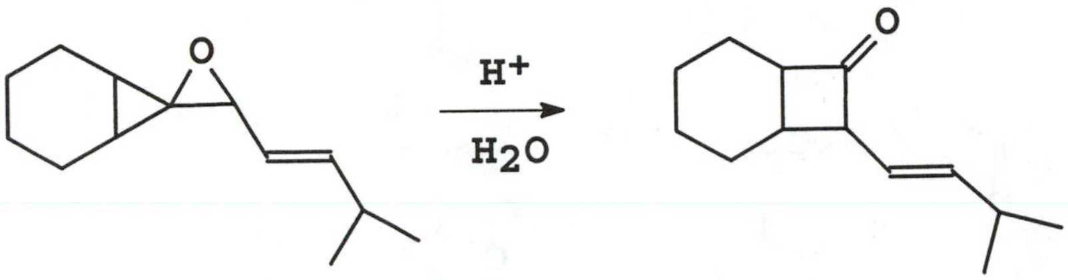

$$\xrightarrow[\text{H}_2\text{O}]{\text{H}^+}$$

22)

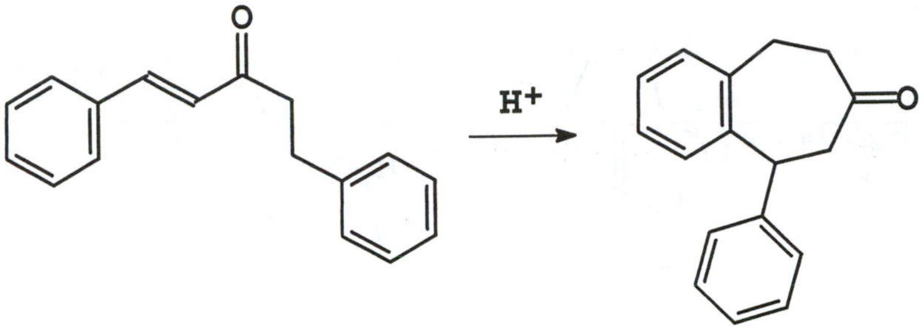

$$\xrightarrow{\text{H}^+}$$

23)

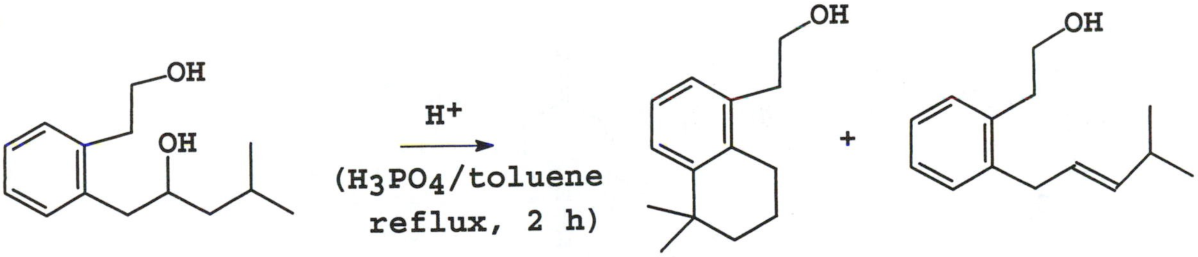

$$\xrightarrow[\substack{(\text{H}_3\text{PO}_4/\text{toluene} \\ \text{reflux, 2 h})}]{\text{H}^+}$$

+

24)

$$\xrightarrow{\text{H}^+}$$

+

59

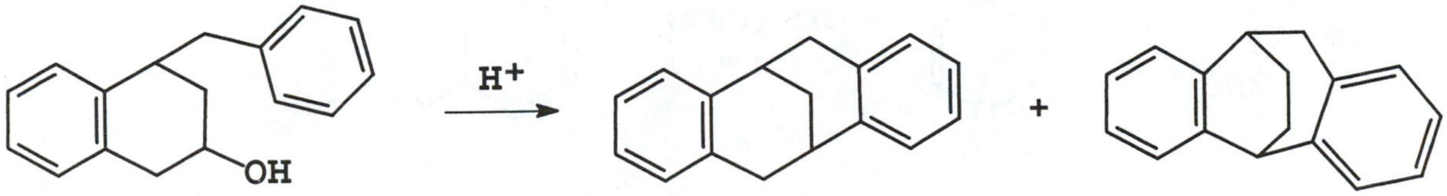

25)

26)

27)

28)

60

29)

30)

31)

32)

33)

34)

ANSWERS TO QUESTIONS 1-34

1)

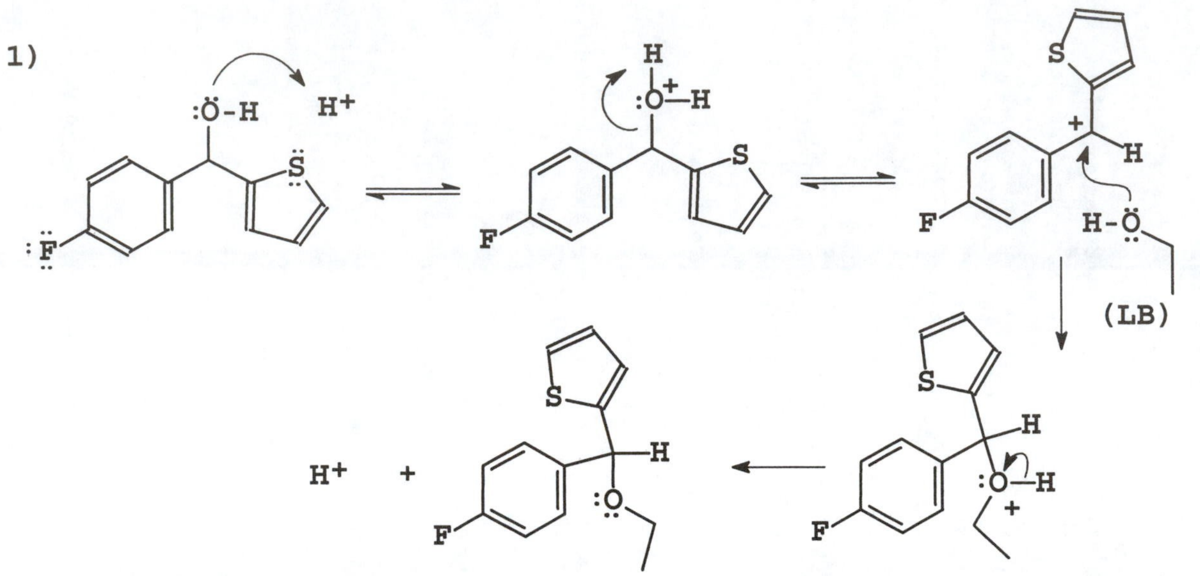

An initial LA/LB reaction leads to the formation of a resonance-
stabilized benzylic carbocation (a Lewis acid), which then reacts
with ethyl alcohol (a Lewis base) to form the product. Note that
(a) non-bonded electrons are more available for donation than
bonded electrons (pi electrons); (b) reaction of the LA
with either aromatic ring would result in the formation of a non-
aromatic species (loss of resonance-stabilization energy) and is
an unproductive process, i.e., would not lead to the formation of
the observed product).

2)

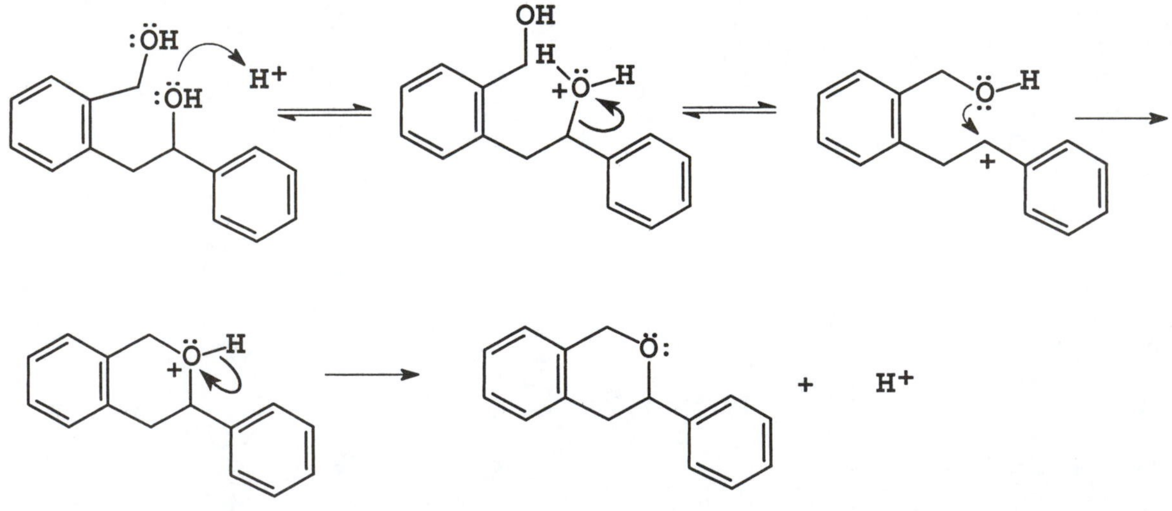

A LA/LB reaction leads to the formation of a benzylic carbocation.
A subsequent intramolecular LA/LB reaction leads to the formation
of the product.

64

3)

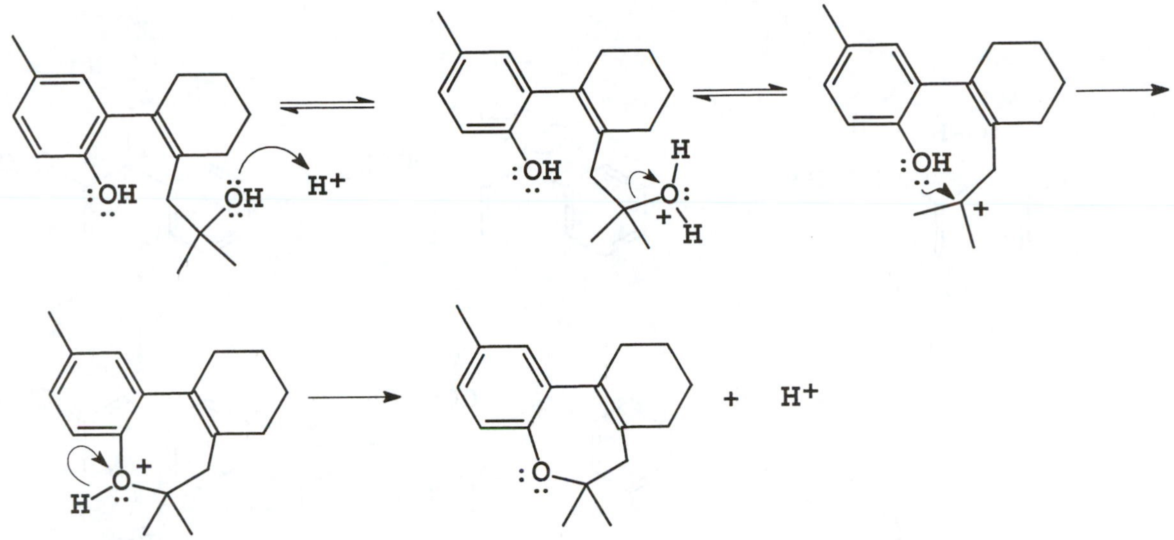

Note that although the phenolic -OH can, in principle, react with the Lewis acid, formation of a phenyl carbocation does not take place because of the low stability of this carbocation. Consequently, only the much more stable tertiary carbocation is formed. This is then followed by an intramolecular LA/LB reaction.

4)

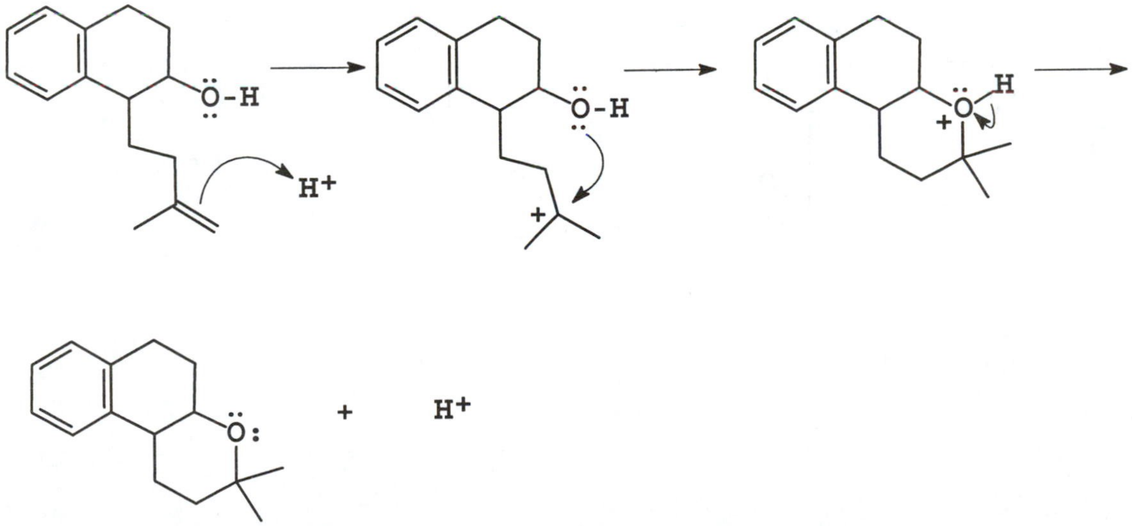

A LA/LB reaction between the pi bond and the hydrogen ion (LA) yields a tertiary carbocation. This is followed by an intramolecular LA/LB reaction, leading to the formation of the observed product.

5)

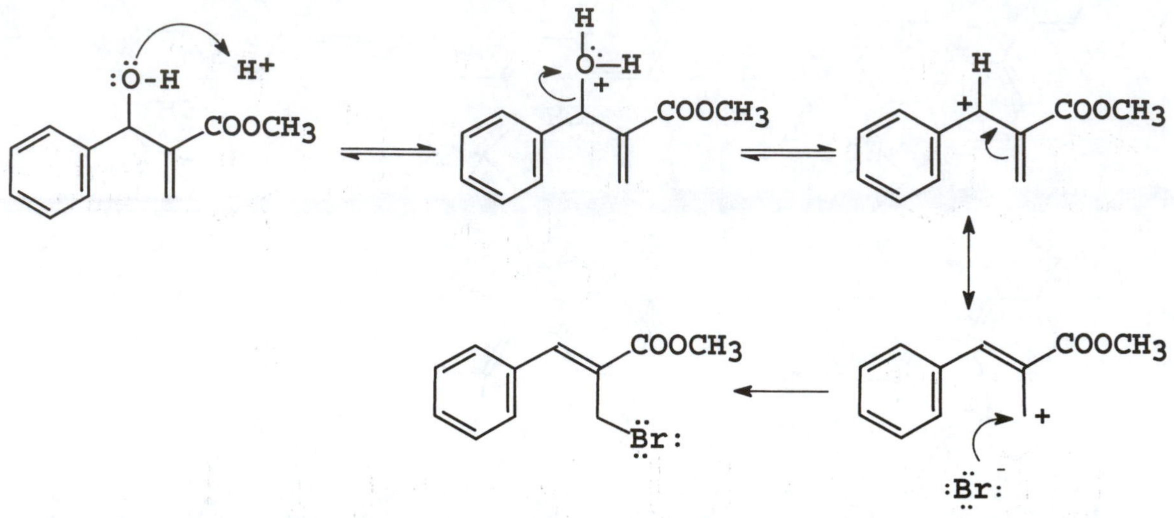

Recall that strong acids are fully-ionized in a solution.
Formation of the resonance-stabilized allylic (and benzylic)
carbocation is followed by a LA/LB reaction to form the
product.

6)

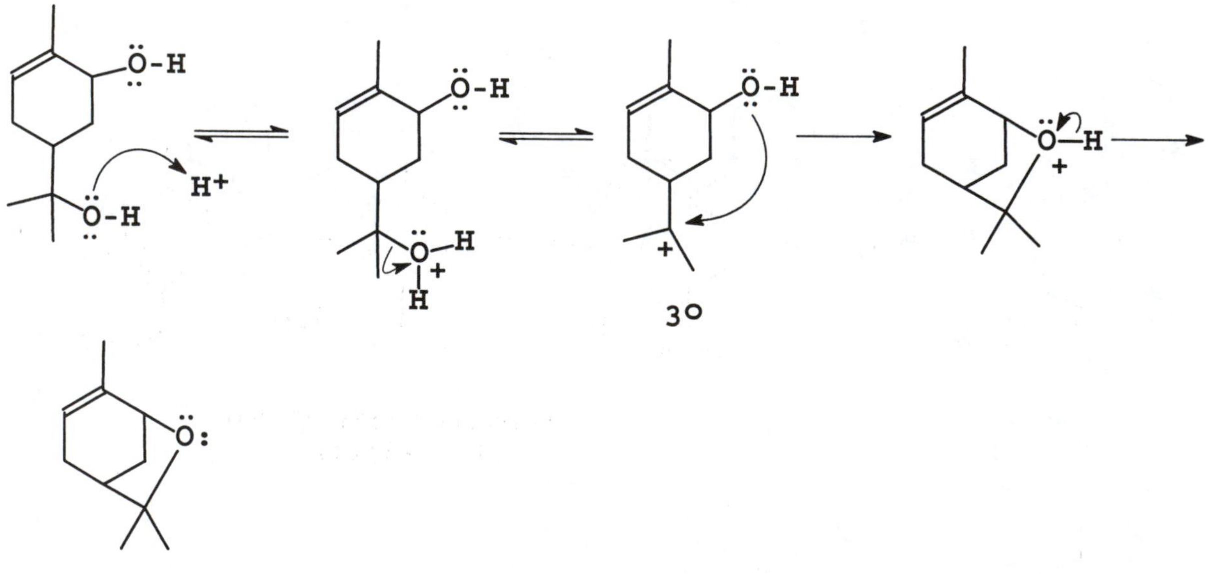

Although either -OH group can react in a LA/LB reaction,
the one that leads to formation of the most stable carbocation
reacts preferentially. A subsequent intramolecular LA/LB
reaction leads to the formation of the product.

7)

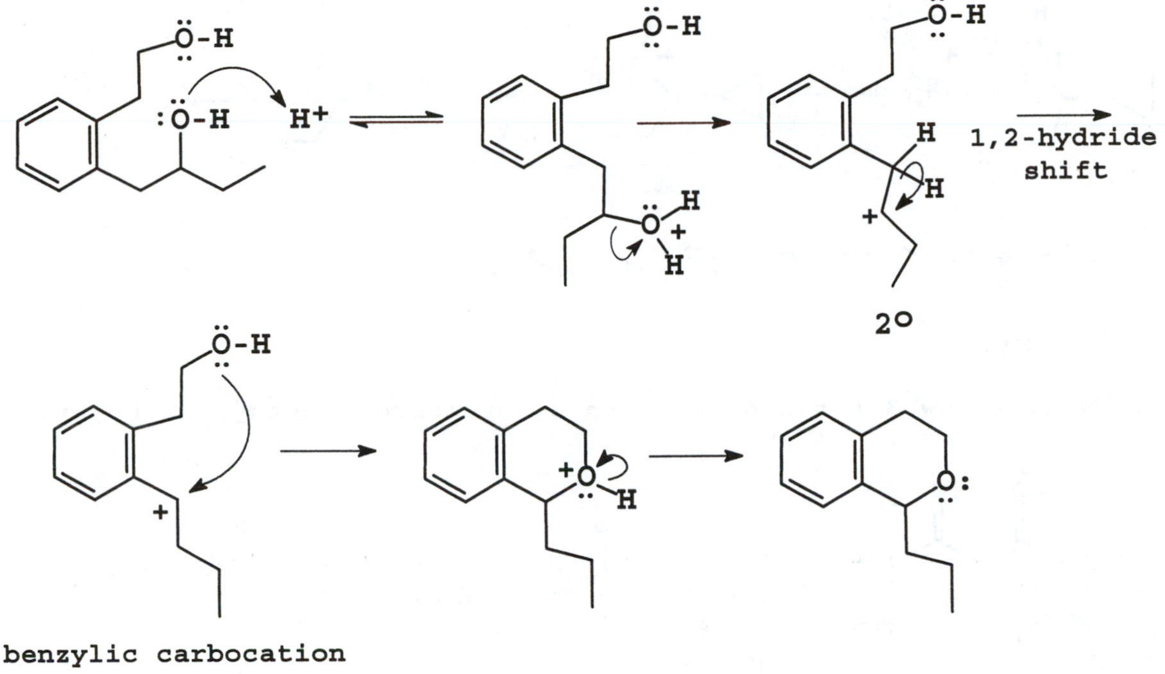

Ö-H
:Ö-H
H+
1,2-hydride
shift
2°
benzylic carbocation

8)

Ö-H
H+
resonance-stabilized
3° & allylic

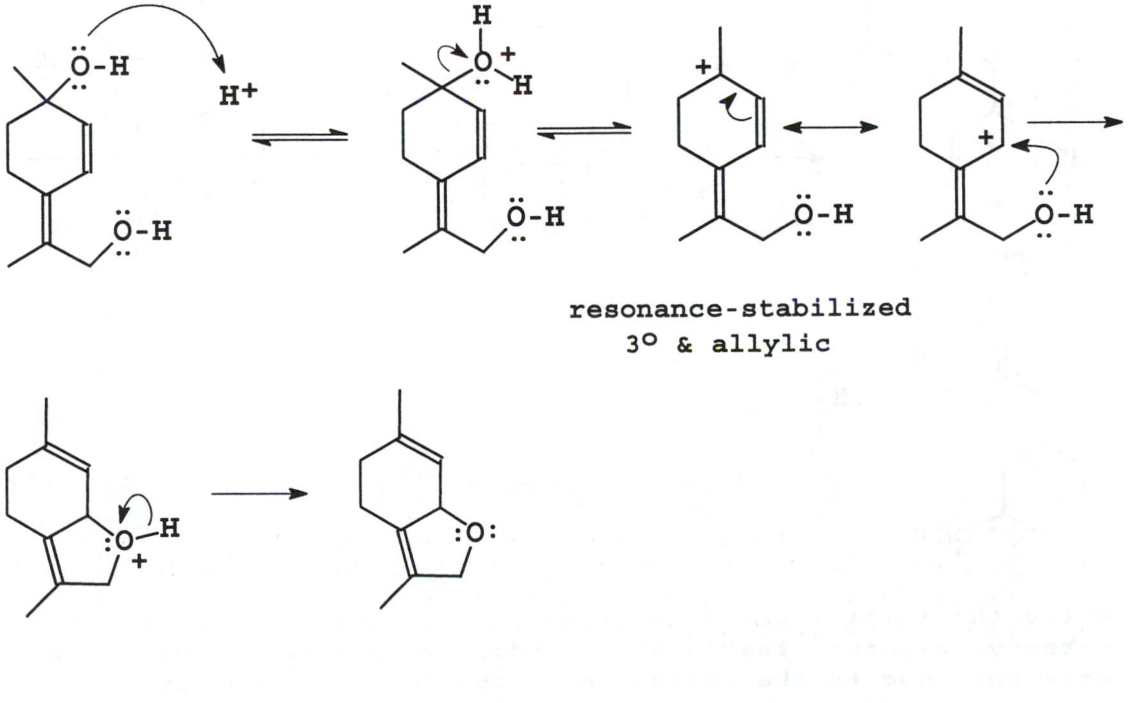

9)

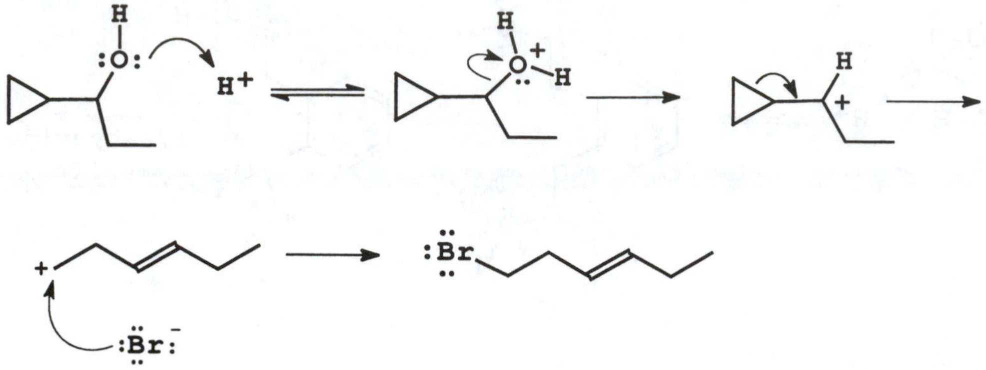

See Minireview 3 for a discussion of cyclopropylmethyl carbocations.

10)

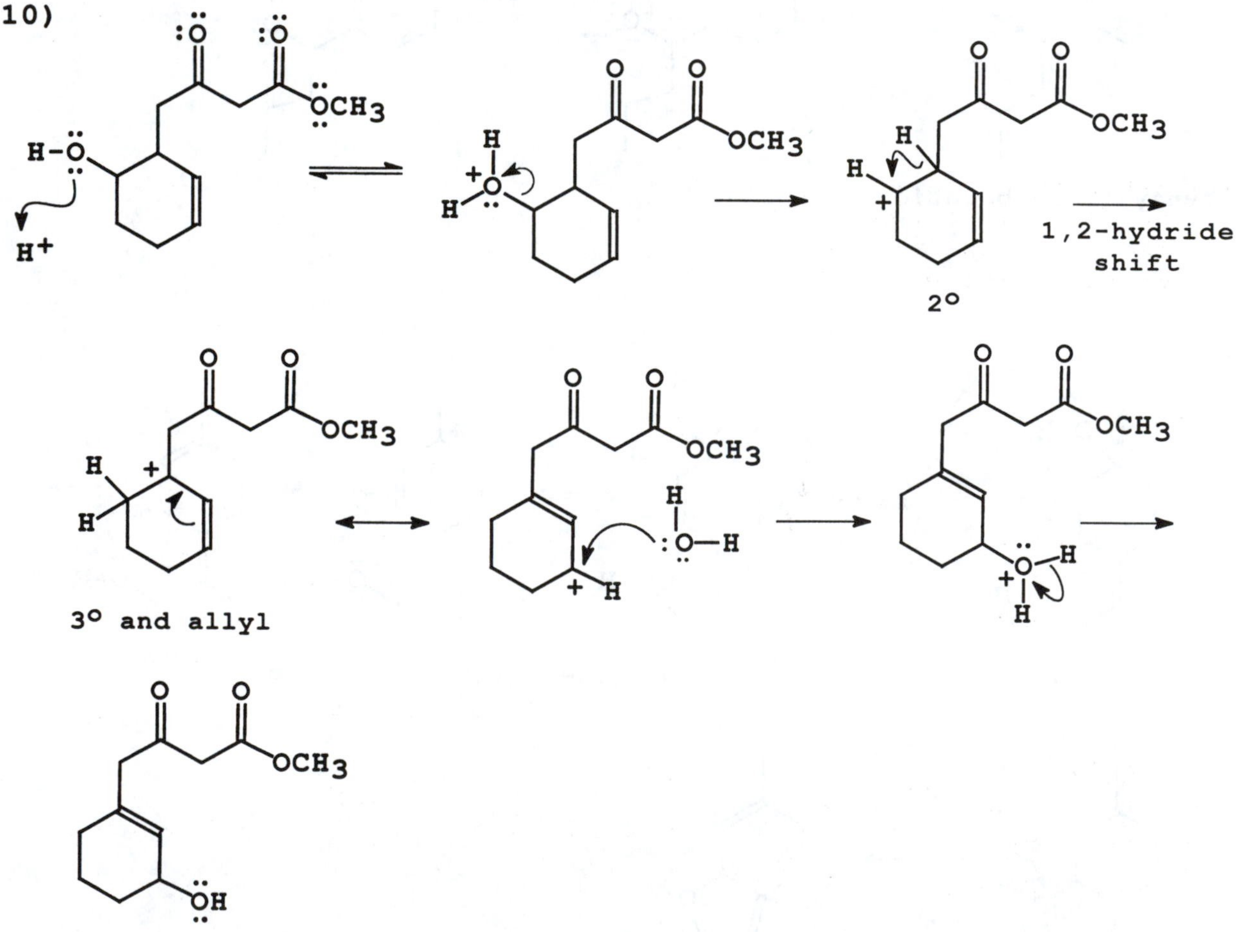

1,2-hydride shift

2°

3° and allyl

While the initial LA/LB reaction can also take place at the carbonyl oxygens, that's an unproductive process, namely, it does not lead to the formation of the observed product.

11)

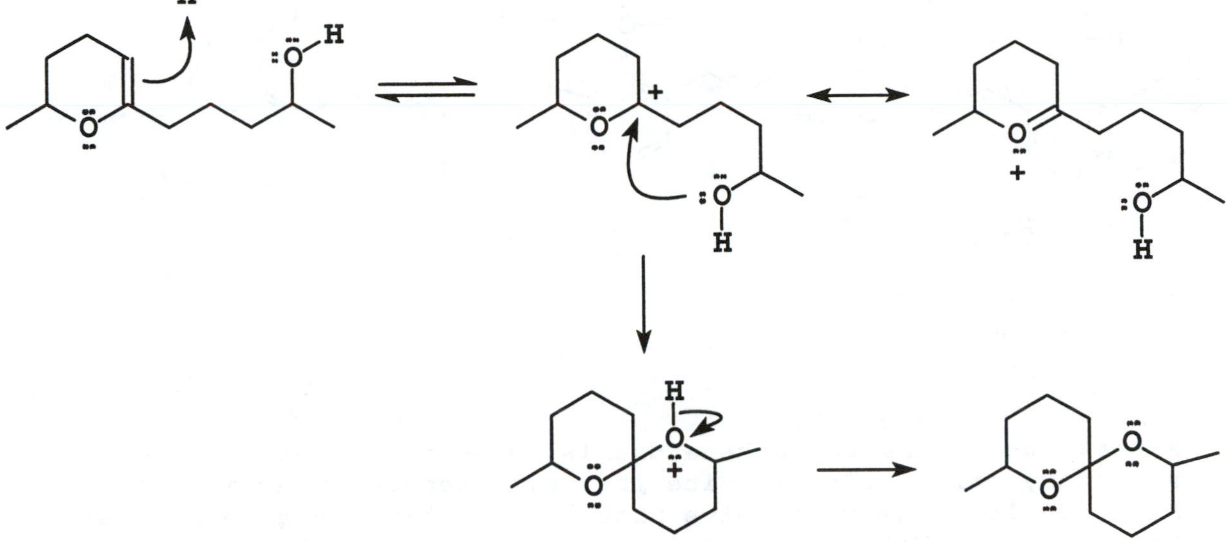

Although the starting material has three functional groups or reactive sites where a LA/LB reaction can take place, the more stable resonance-stabilized cation is formed preferentially.

12)

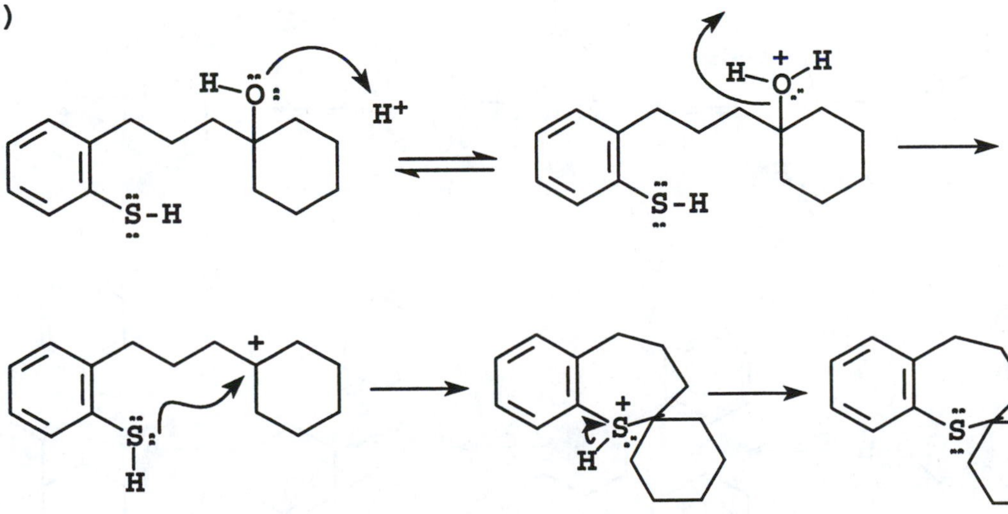

13)

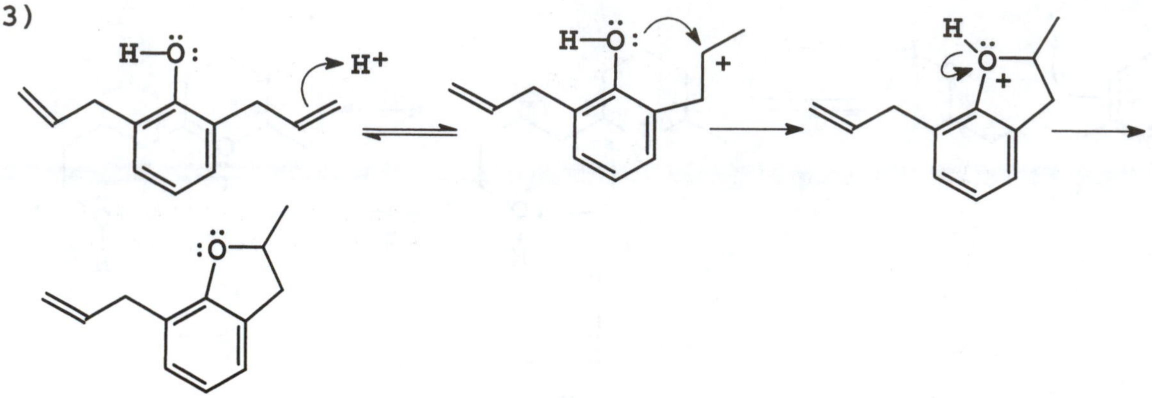

Either one of the two alkene double bonds can react to form a
secondary carbocation (notice that simultaneous reaction of
both double bonds would give rise to a highly unstable species
having two positive charges, therefore, it never happens). As
noted earlier, while the aromatic ring can react in a LA/LB
reaction, this is energetically unfavorable (loss of aromaticity).

14)

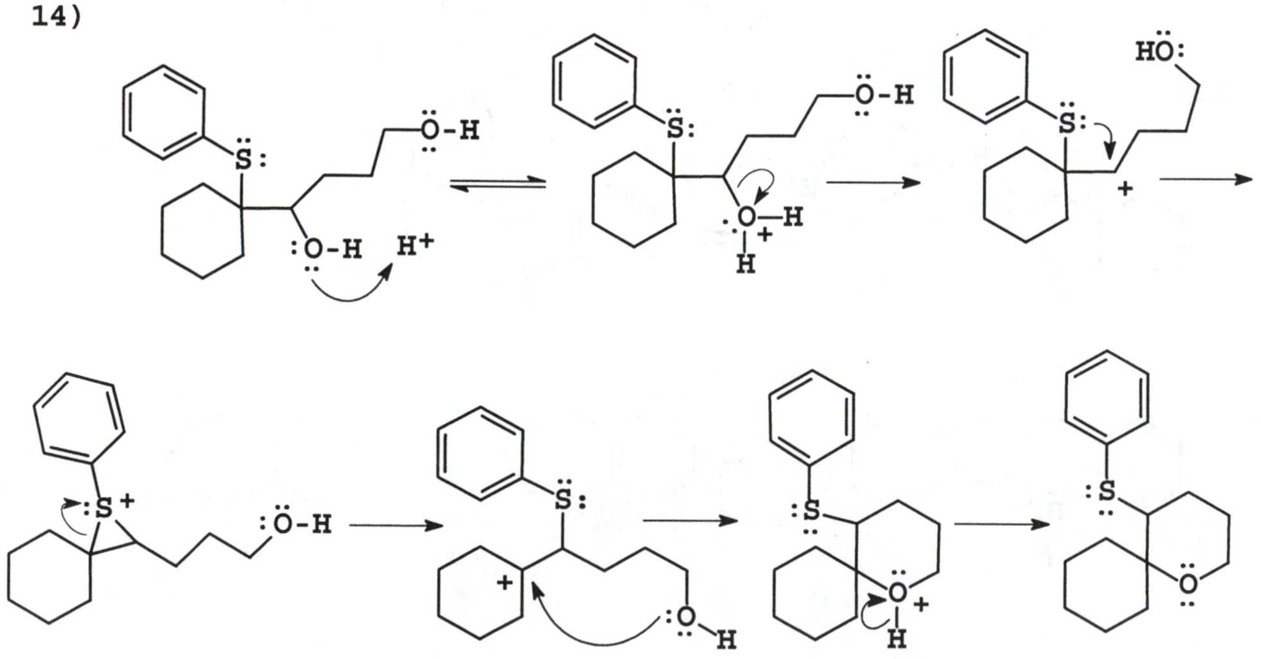

15)

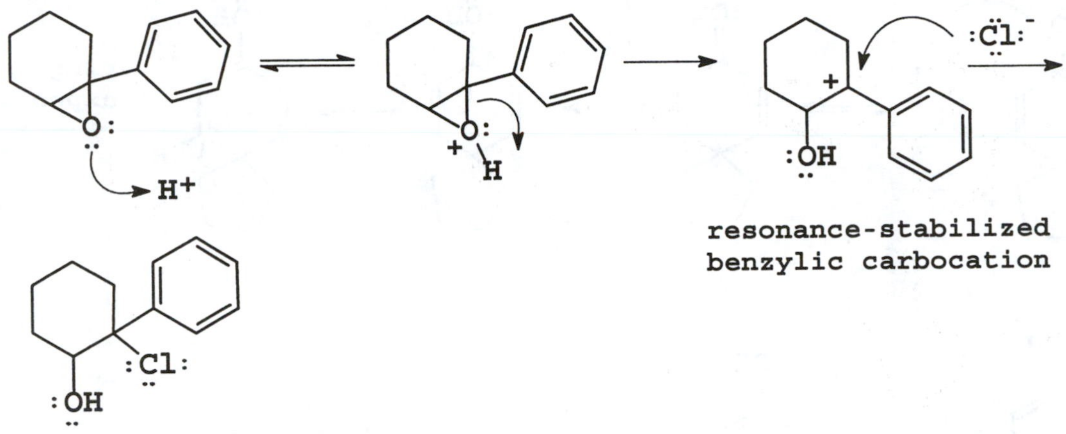

resonance-stabilized
benzylic carbocation

A LA/LB reaction is followed by ring-opening to yield the most stable carbocation (benzylic instead of the less stable secondary carbocation). A second LA/LB reaction yields the observed product.

16)

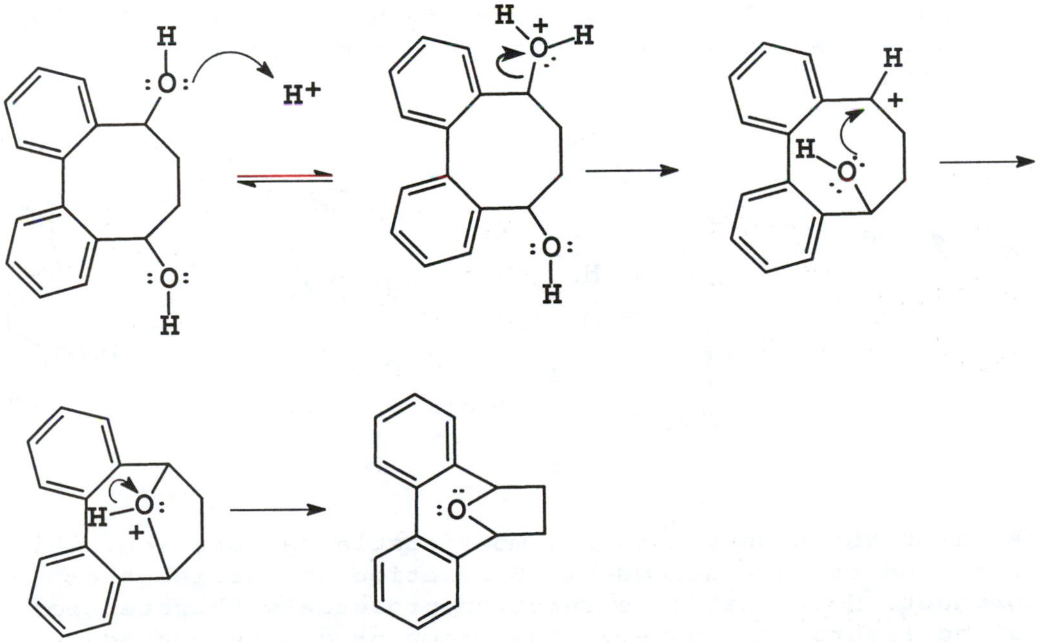

17)

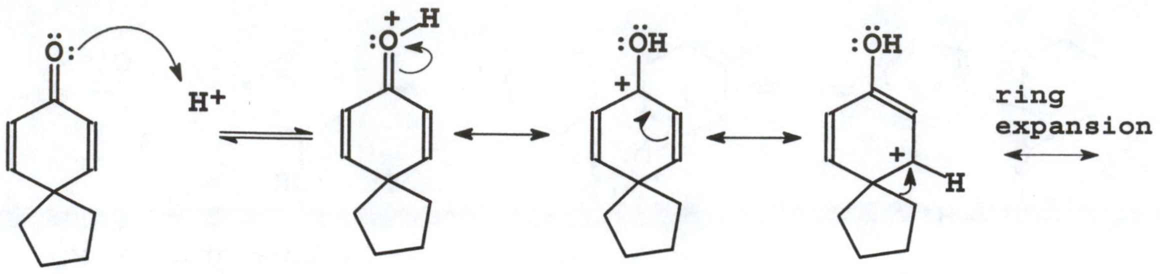

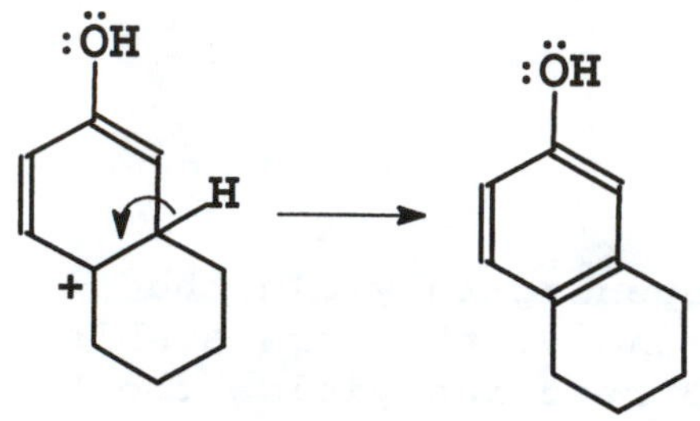

The driving force for this reaction is primarily due to the formation of the aromatic ring. Recall that aromatic compounds are cyclic, planar, satisfy Huckel's 4n + 2 rule, where n is the number of pi electrons and, can sustain a ring current.

18)

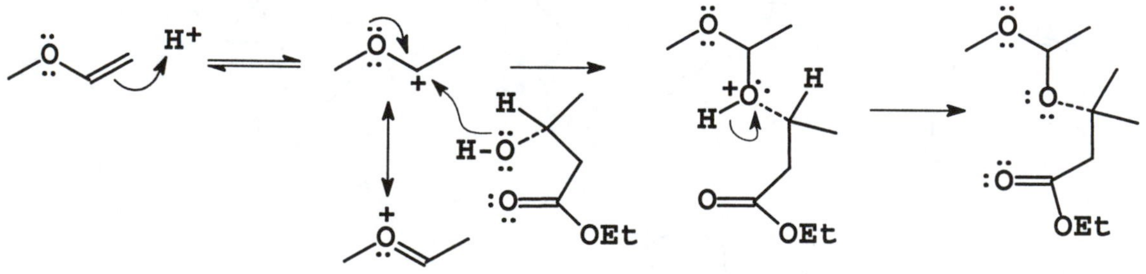

A LA/LB reaction yields the most stable carbocation, which then reacts in a second LA/LB reaction to furnish the product. Note that this reaction proceeds with retention of configuration, namely, the atoms or groups bonded directly to the stereogenic (chiral) center are oriented in space the same way in the product, as in the starting material, since none of the bonds to the stereogenic center is broken during the course of the reaction.

19)

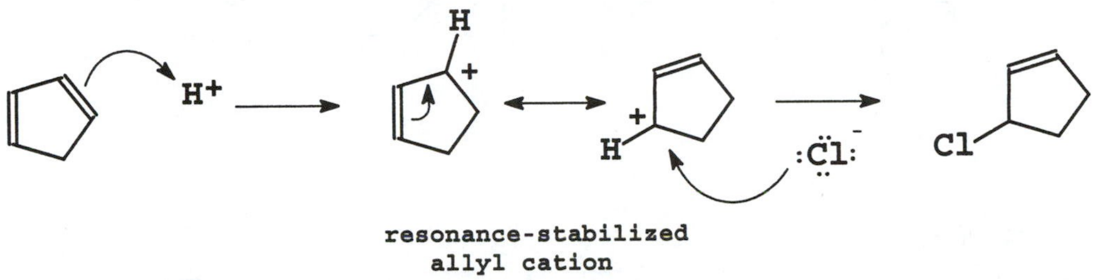

resonance-stabilized
allyl cation

20)

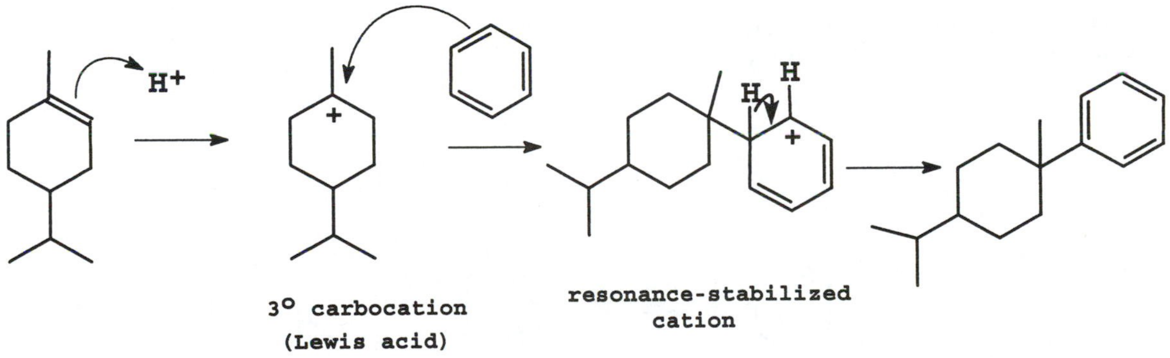

3° carbocation
(Lewis acid)

resonance-stabilized
cation

A LA/LB reaction leads to the formation of the most stable carbocation. This is followed by electrophilic aromatic substitution reaction, specifically, a Friedel-Craft's alkylation reaction (see Minireview 3).

21)

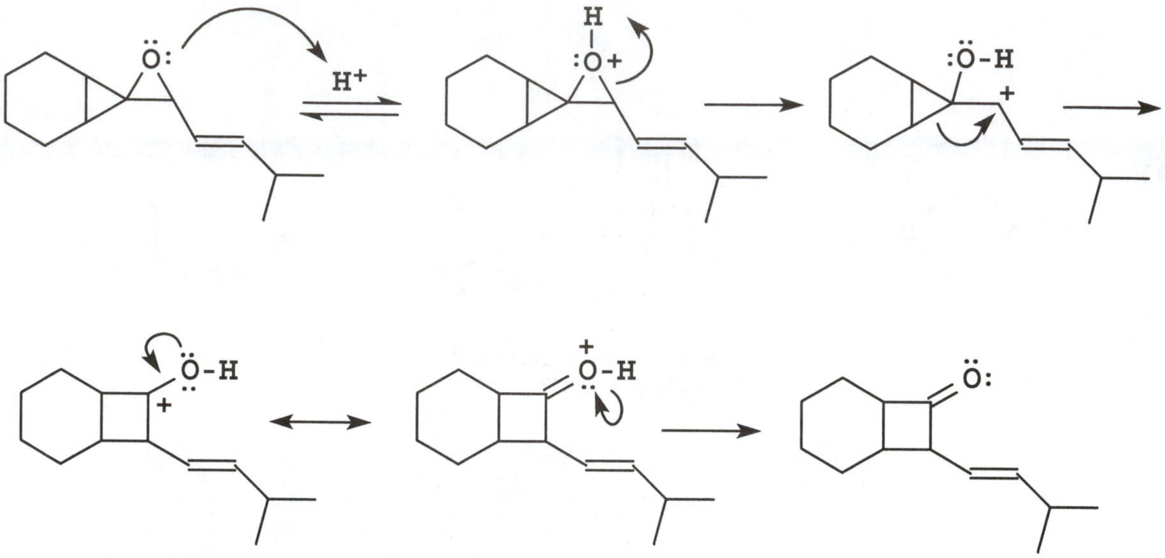

A LA/LB reaction is followed by ring opening of the epoxide (oxirane) ring to form a resonance-stabilized allyl carbocation. Ring expansion results in the relief of ring strain and the formation of a resonance-stabilized carbocation.

22)

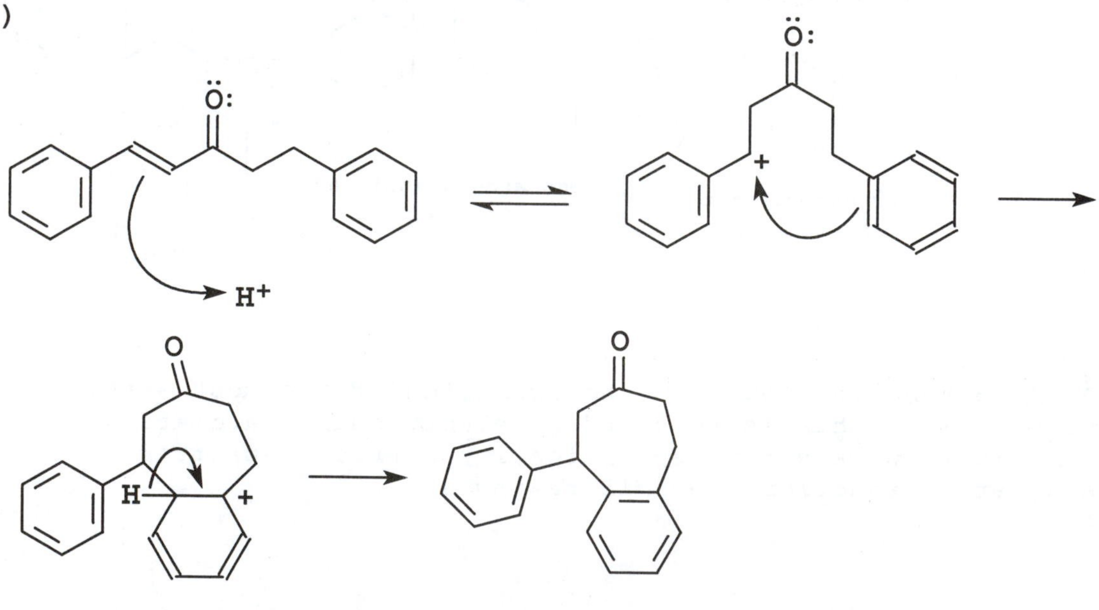

An initial LA/LB reaction yields the most stable carbocation (note that carbocations with a positive charge on a carbon alpha to a carbonyl group are highly unstable). This is followed by an intramolecular Friedel-Craft alkylation reaction.

23)

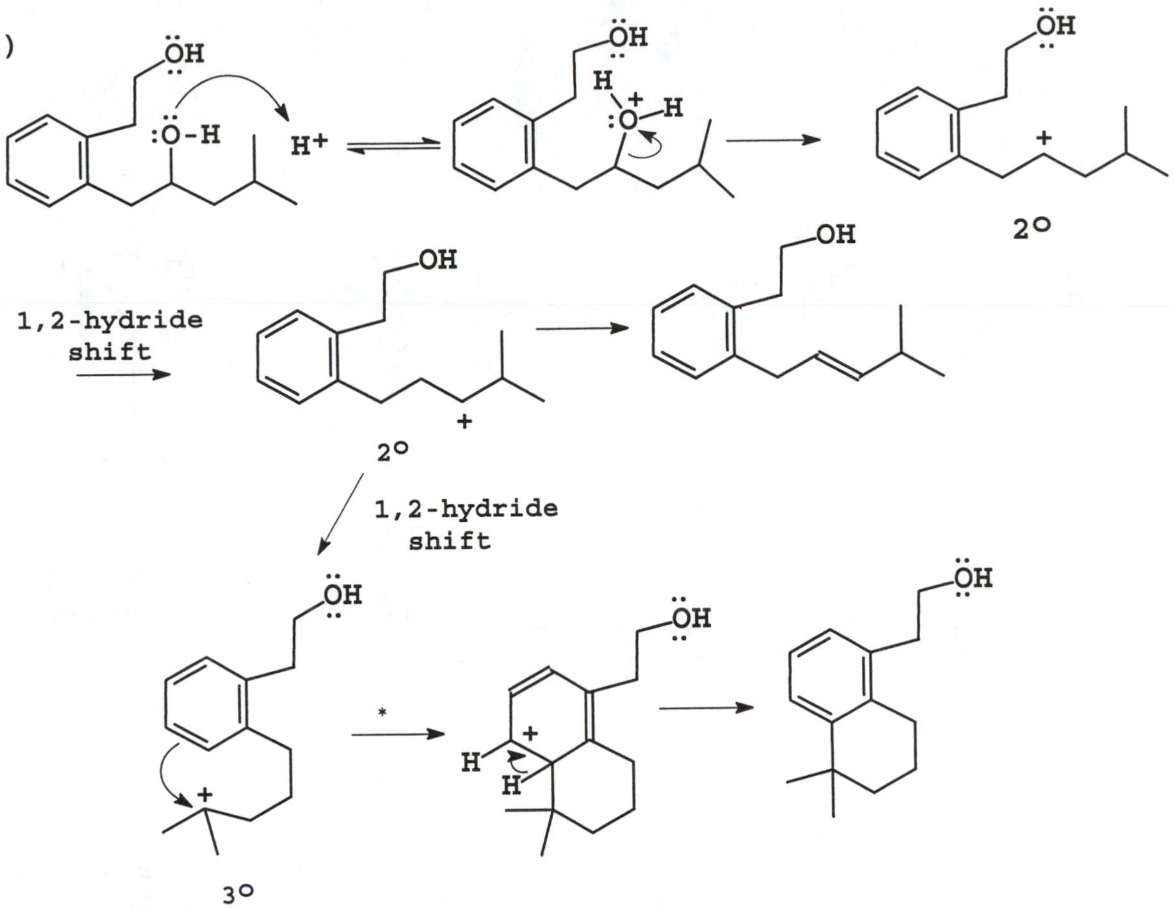

H+
2°
1,2-hydride
shift
2°
1,2-hydride
shift
3°
*

*Intramolecular Friedel-Craft alkylation reaction

24)

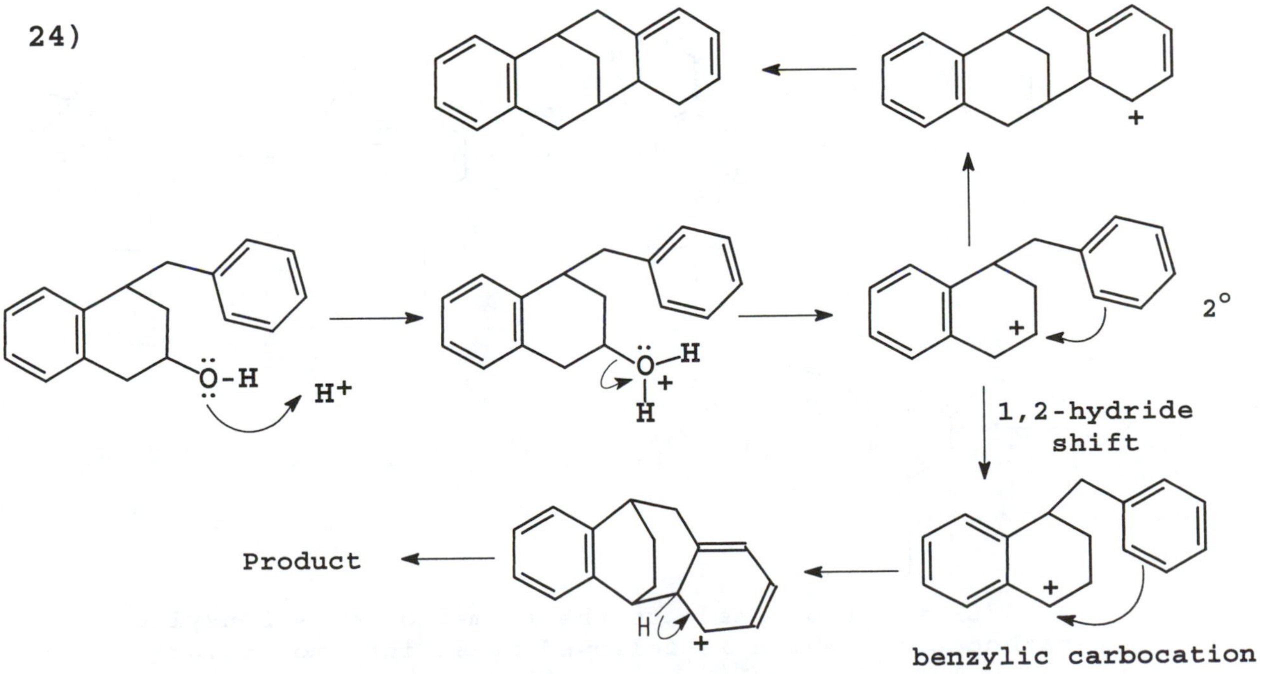

H+
2°
1,2-hydride
shift
benzylic carbocation
Product

25)

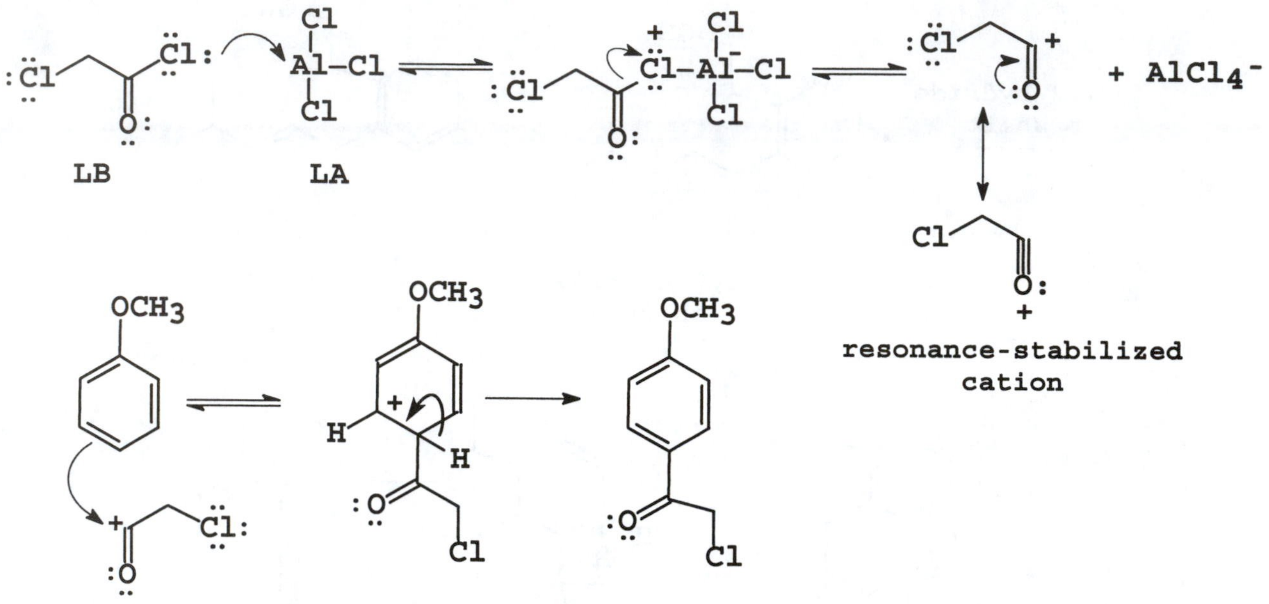

This is an example of the Friedel-Craft acylation reaction (see Minireview 3). Since the methoxy group is an activating o and p-directing group, the major product formed is that shown above.

26)

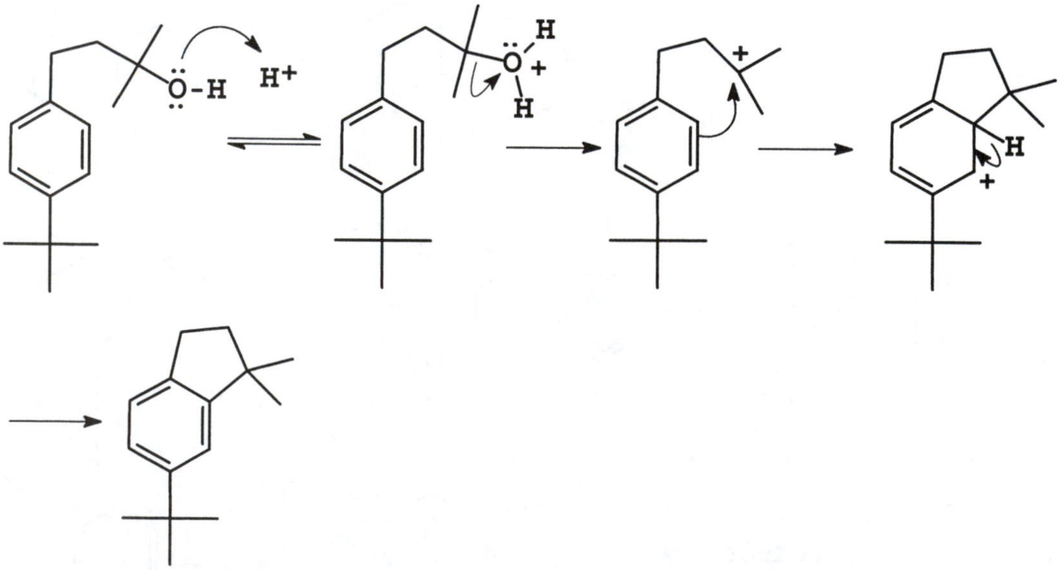

A LA/LB reaction leads to the formation of a benzylic carbocation, which is followed by an intramolecular Friedel-Craft alkylation reaction.

27)

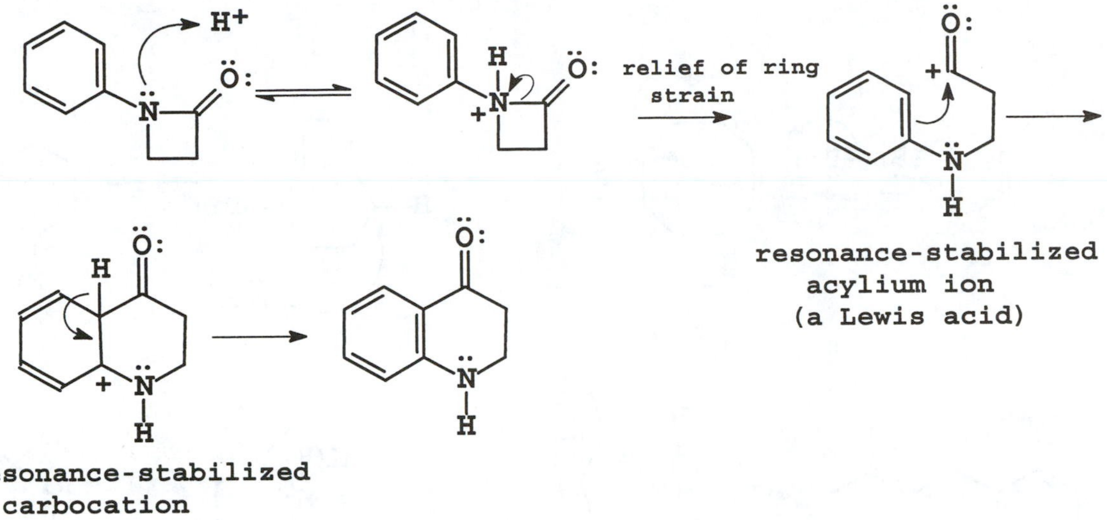

H+
relief of ring strain
resonance-stabilized acylium ion (a Lewis acid)
resonance-stabilized carbocation

The pair of non-bonded electrons in the starting material is more available for donation than in acyclic amides. Why?

28)

H+

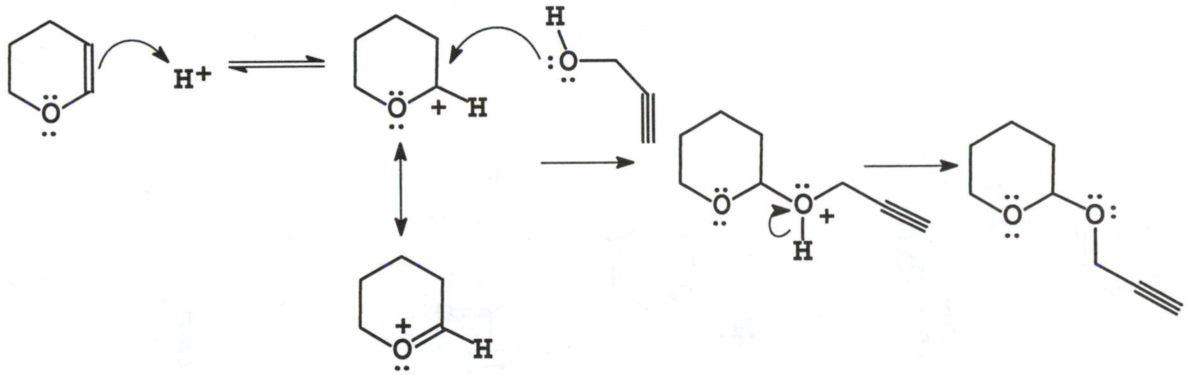

Two consecutive LA/LB reactions lead to the formation of the product. Note that a LA/LB reaction at the oxygen atom of the starting material is an unproductive process.

H+

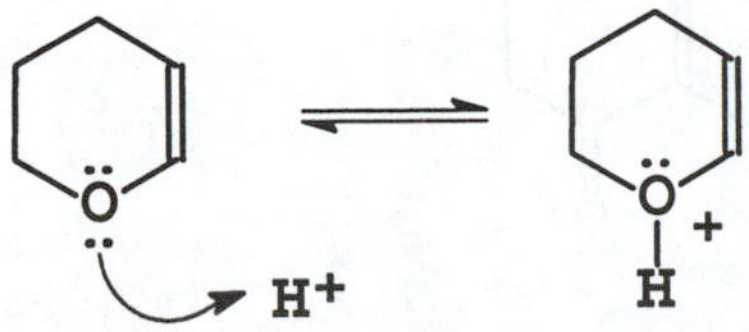

29)

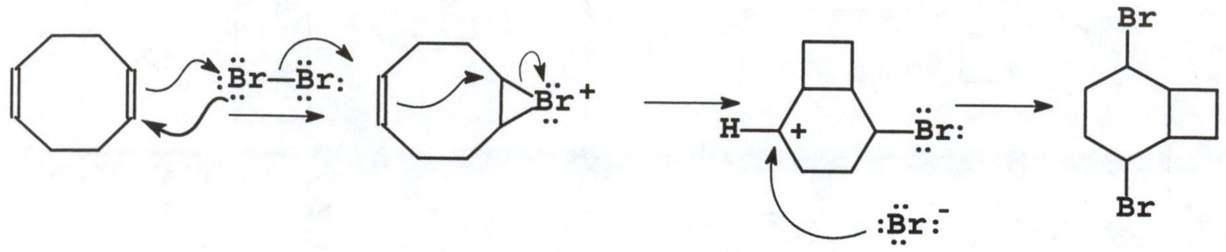

30)

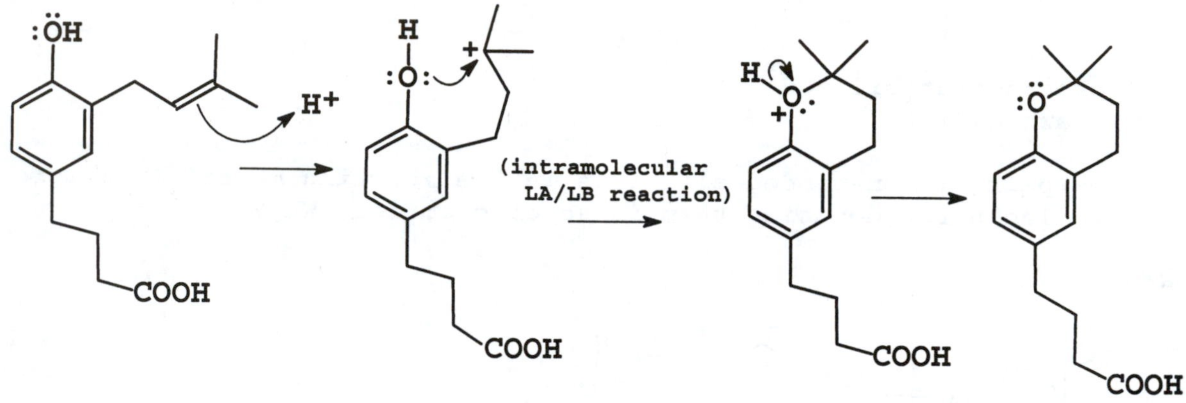

(intramolecular LA/LB reaction)

31)

LB   LA

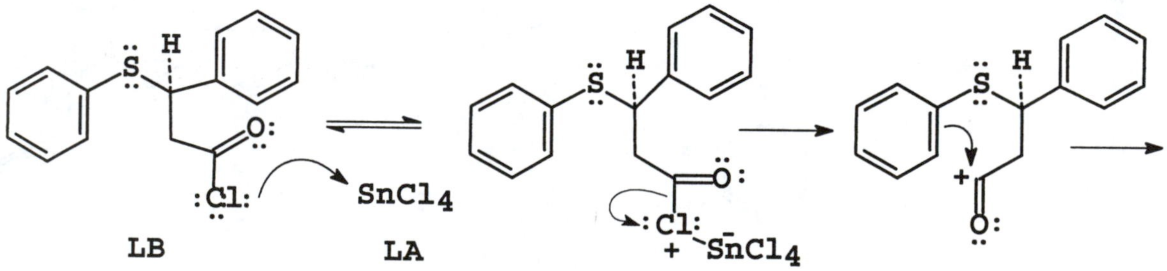

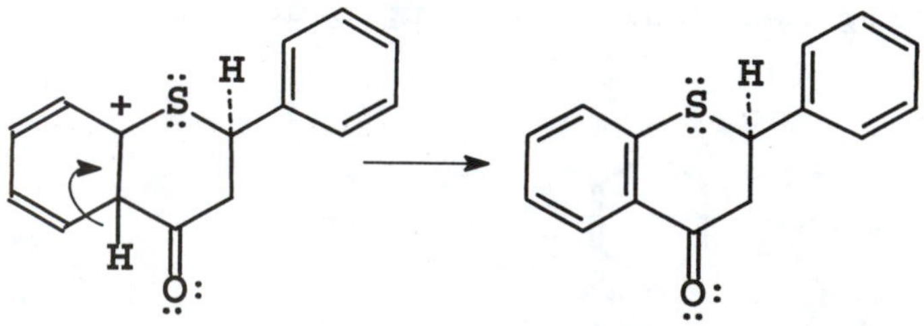

Reaction proceeds with retention of configuration, since none of the bonds to the stereogenic center have been broken during the course of the reaction.

32)

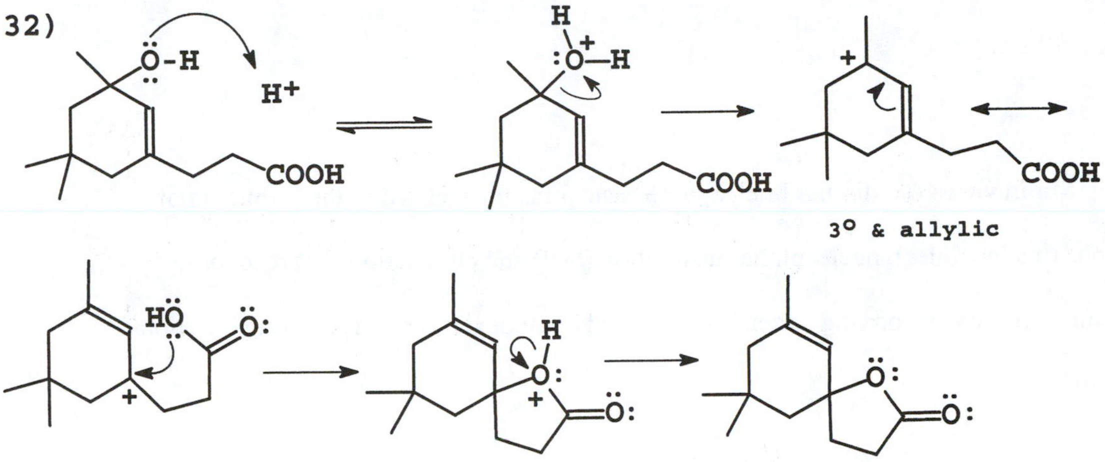

H+
COOH
3° & allylic

33)

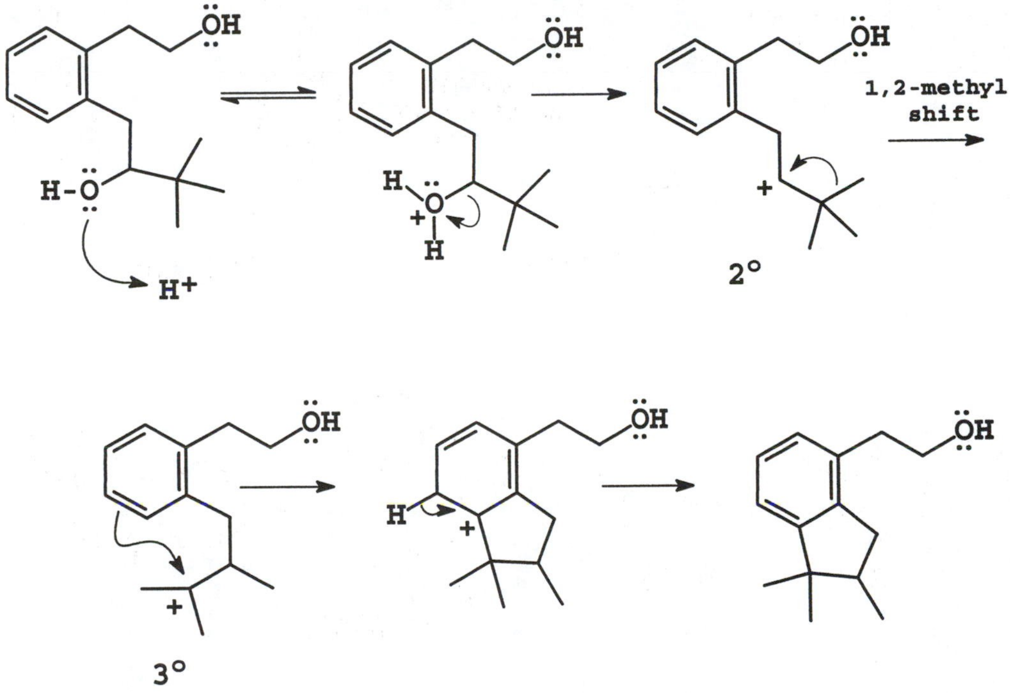

H+
2°
1,2-methyl shift
3°

34)

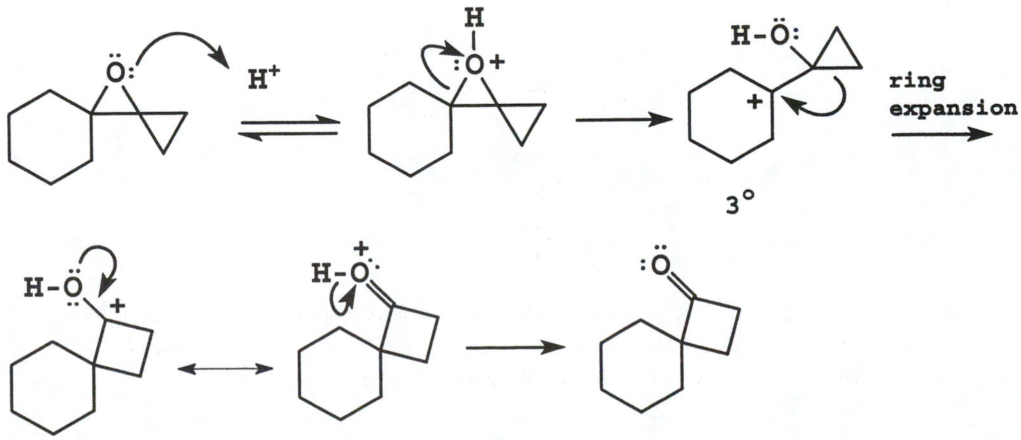

H+
3°
ring expansion

Minireviews 4-6 discuss briefly some basic principles related to the formation of anions (nucleophiles), nucleophilic substitution (S_N2) and elimination (E_2) reactions, chemical reactivity and ring strain. These should be studied prior to attempting questions 35-50.

MINIREVIEW 4 Formation of Anions (Nucleophiles)

Introduction Before discussing anions (Lewis bases or nucleophiles), it is helpful to
remember that the chemistry of alkyl halides, aldehydes and ketones, and carboxylic acid
derivatives (acid chlorides, anhydrides, esters, and amides) can be simply described and
best understood by equations (1-3) below, without recourse to memorization. Thus, the
typical reaction of alkyl halides is nucleophilic substitution (S_N2) (eq 1) (see Minireviews
5 and 6 for more details and some variations of the same theme), the typical reaction of
aldehydes and ketones is nucleophilic addition (eq 2) (Minireview 7), and the typical
reaction of carboxylic acid derivatives is nucleophilic acyl substitution (eq 3)
(Minireview 8). These three general reactions can be viewed as Lewis base/Lewis acid
reactions and are discussed in greater detail in Minireviews 5-8. Reactions 1-3 proceed
most readily when the Lewis base (nucleophile) has a negative charge. Thus, *anions are
typically generated for subsequent use in reactions of the type shown below.*

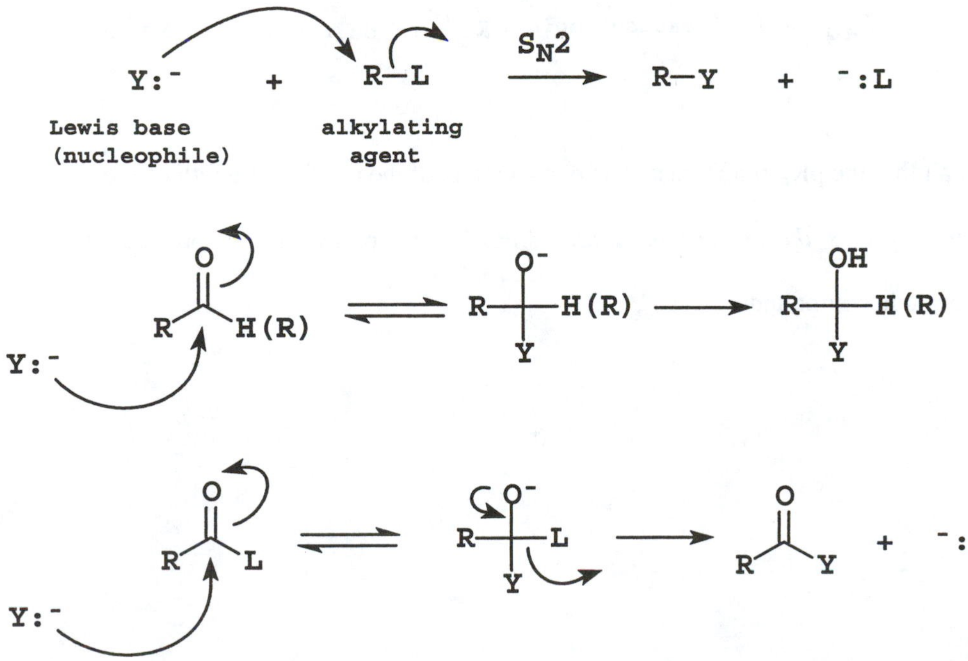

$$Y:^- \quad + \quad R-L \quad \xrightarrow{S_N2} \quad R-Y \quad + \quad ^-:L \qquad \text{eq 1}$$

Lewis base   alkylating
(nucleophile)   agent

eq 2

eq 3

81

Factors to consider prior to generating an anion include the relative acidity of the starting material (HA) and the type of base (B⁻) to use. It can be readily shown that the equilibrium constant (K_{eq}) for the general acid-base reaction shown below (equation 4) is given by equation 5, where K_a is the acidity constant of the reactant acid (HA) and K_a' is the acidity constant of the product acid (HB). Since the pK_a's of a large number of Bronsted acids have been determined and are readily available, and since $pK_a = - \log K_a$, an approximate K_{eq} for any acid-base reaction can be readily calculated using equation 5 (vide infra). In order for reactions (1-3) above to proceed readily, a high concentration of the anion (nucleophile) is desirable. Hence, *a sufficiently strong base is typically selected to generate (Y) rapidly and quantitatively.*

$$\text{HA} \;+\; \text{B}^- \;\rightleftharpoons\; \text{HB} \;+\; \text{A}^- \qquad \text{eq 4}$$
reactant acid   product acid

$$K_{eq} \;=\; K_a\,(\text{reactant acid}) \,/\, K_a'\,(\text{product acid}) \qquad \text{eq 5}$$

Recall that the pK_a of a Bronsted acid is an index of the relative strength of the acid. *The lower the pK_a is, the stronger the acid.* Table 1 lists the pK_a values of some common types of organic compounds.

TABLE 1. pK_a Values of Some Common

Classes of Organic Compounds*

	pK_a
RSO_3H	-6.5
RCOOH	4-6
ArSH	6-8
RSH	10-11
ArOH	8-11
ROH	16-18
$RCONH_2$	17
RCH_2CHO	19-20
RCH_2COR	19-20
RCH_2COOR	25

*values are approximate

The acidity of a substance is greatly affected by the presence of one or more functional groups capable of delocalizing the negative charge (via resonance) in the corresponding anion (Minireview 2). Common functional groups that enhance acidity include NO_2, COR, CN, COOR, SO_2R and phenyl (Ph). These groups differ in their ability to delocalize the negative charge in the anion, hence the extent to which acidity is affected is dependent on the *nature* and *number* of such functional groups. For example, in a series of compounds of the type RCH_2X, the pK_a varies as follows: $X = NO_2$ (10), (C=O)R (20), CN (25), COOR (25) and SO_2R (29). The presence of two such groups reduces the pK_a by about one half.

Selection of Base

In selecting a base to be used in the generation of an anion, the *strength* of the base and its *compatibility* with any other functional groups that might present in the molecule, are factors of paramount importance. In general, a sufficiently strong base is selected so that the desired anion is generated rapidly and quantitatively. Secondly, depending on the structure of the acidic molecule (HA), the base used may have to be *non-nucleophilic* (vide infra). Common bases used to generate anions include the following:

(a) Alkoxide ions ($RO^- M^+$, where $M^+ = Na^+$ or K^+

Alkoxide ions are conveniently generated by reacting dry methanol ($pK_a \sim 15.5$), ethanol ($pK_a \sim 16$), or t-butyl alcohol ($pK_a \sim 18$) with sodium or potassium metal under a nitrogen atmosphere. The generated base is then used in situ (in the flask, without isolation). *Note that alkoxides can function as both bases and nucleophiles.*

$$EtOO-CH_2-COOEt \;+\; CH_3CH_2O^- \;\rightleftharpoons\; EtOO-\ddot{C}H-COOEt \;+\; CH_3CH_2OH$$

$$pK_a \; 13 \qquad\qquad pK_a \; 16$$

Using equation (5) above, $K_{eq} = 10^{-13}/10^{-16} = 10^3$, therefore the formation of the anion (:Y^-) is greatly favored (equilibrium lies to the right).

In cases where a reactant having more than one acidic hydrogen is treated with base, *the most acidic hydrogen will react first*, since it leads to the formation of the more stable anion. The following examples illustrate this concept.

(a) Amines

Recall that basicity in amines follows the order

$$3° > 2° > 1° > NH_3 > Ar\text{-}NH_2$$

increasing base strength

Triethylamine (TEA), pyridine and N-methylmorpholine (NMM) are organic bases

that are widely used in organic synthesis.

When the objective is to generate an anion via an acid-base reaction, or to induce an

E$_2$ elimination reaction (vide infra), the use of a sterically-hindered, non-nucleophilic

base, such as diisopropylethyl amine (DIEA), 1,8-diazabicyclo[5.4.0]undecen-7-ene

(DBU), or 1,5-diazabicyclco[4.3.0]non-5-ene (DBN) should be considered.

DIEA   **DBU**   **DBN**

If a much stronger, non-nucleophilic base is needed, then use of one of the following bases is indicated:

Sodium or potassium hydride NaH or KH   (:H⁻)

Lithium diisopropyl amide (LDA)

(lithium diisopropyl amide can be generated *in situ* by treating diisopropyl amine with n-butyl lithium. The pK_a of diisopropyl amine is ~36).

(b) Alkyl lithium reagents (R:⁻ Li⁺)

Alkyl lithium bases, such as methyl lithium, n-butyl lithium, and t-butyl lithium, are *strongly basic* and *nucleophilic*. Because these alkyl lithium reagents are the conjugate bases of the corresponding hydrocarbons ($pK_a \sim 50$), their high basicity and reactivity requires the use of an inert atmosphere and anhydrous conditions. Recall that the lower the pK_a is, the stronger the acid and the weaker the conjugate base. Conversely, the higher the pK_a is, the weaker the acid and the stronger the conjugate base.

(d) Miscellaneous Bases

These include bases such as sodium or potassium bicarbonate, carbonate or hydroxide. The pK_a's of carbonic acid, bicarbonate and water are 6.4, 10.3 and 15.7, respectively. Consequently, base strength follows the order $HO^- > CO_3^{2-} > HCO_3^-$.

Examples illustrating the *rationale* underlying the selection of a particular base are given below. See Minireview 5 for additional examples. It would be very beneficial to consider the logic behind the use of a particular base, or reaction conditions, as you work through the problems in this workbook.

Example 1 The following transformation was successfully achieved by using a bulky and non-nucleophilic base, greatly minimizing competing S_N2 and ring-opening reactions between the base and the highly reactive functionalities in the starting material and product.

Example 2 The objective was to induce E_2 elimination only, leading to the formation of the desired product. The selection of base was critical for the success of this reaction because of the high reactivity of the product. The latter is an allylic bromide (highly reactive in S_N2 reactions), thereby necessitating the use of a bulky and non-nucleophilic base.

MINIREVIEW 5 Nucleophilic Substitution (S_N2) and Elimination Reactions (E_2)

A. S_N2 Reactions

The S_N2 reaction is *a concerted, one-step reaction involving backside attack by a strong Lewis base (nucleophile) on an sp^3-hybridized carbon bearing a partial positive charge ($\delta+$) (Lewis acid) and a leaving group (L)*. The reaction proceeds with *inversion of configuration*. S_N2 reactions are greatly influenced by the nature of the alkylating agent (R-L), nucleophile (:Y⁻), leaving group (L), and solvent.

$$Y:^- \quad + \quad R{-}L \quad \longrightarrow \quad R{-}Y \quad + \quad ^-:L$$

nucleophile
(Lewis base)

(1) Alkylating Agent (R-L)

The order of reactivity of halides in S_N2 reactions is

```
benzyl halides > CH₃X        > 1° > 2° > 3° > vinyl halides
            allyl halides                    aryl halides
                                             (unreactive)

                        ⟵
                  increasing reactivity
```

NOTES (a) Tertiary alkyl halides undergo E_2 reactions, while vinyl and aryl halides do *not* react. Recall that aryl halides bearing electron-withdrawing groups undergo *nucleophilic aromatic substitution* reactions. (b) The reactivity profile of α-haloesters, α-haloethers and α-haloketones is similar to that of benzylic halides.

(2) Strength and Nature of Lewis Base (Nucleophile)

(a) In general, *nucleophilicity increases with increasing basicity, provided nucleophiles with a common nucleophilic atom are compared.* Thus, a nucleophile with a negative charge (:Y⁻) will always be more basic, and therefore more nucleophilic than a neutral nucleophile (Y:). With the exception of amine and phosphorus nucleophiles, nucleophiles used in S_N2 reactions are anions, namely, they have a negative charge (:Y⁻). As pointed out earlier, the lower the pK_a is, the stronger the acid and the weaker the conjugate base.

$$R—O^- \quad > \quad HO^- \quad > \quad Ar\text{-}O^- \quad > \quad \underset{R}{\overset{O}{\underset{}{\|}}} C\text{-}O^-$$

increasing base strength ←

(b) The <u>size</u> of the atom that donates the pair of electrons has a profound effect on its nucleophilicity. *The larger the atom is, the greater its nucleophilicity.* Thus,

$$:PH_3 \quad > \quad :NH_3$$

$$RSe^- > R—S^- > R—O^-$$

$$I^- > Br^- > Cl^- > F^-$$

increasing size ←

increasing nucleophilicity ←

(c) Nucleophiles that are capable of reacting at more than one atom, thereby giving rise to two different products, are referred to as *ambident nucleophiles.* The site of reaction in ambident nucleophiles is primarily determined by the nature of the solvent used in an S_N2 reaction, and the size of the atom. Some examples of ambident anions are shown below.

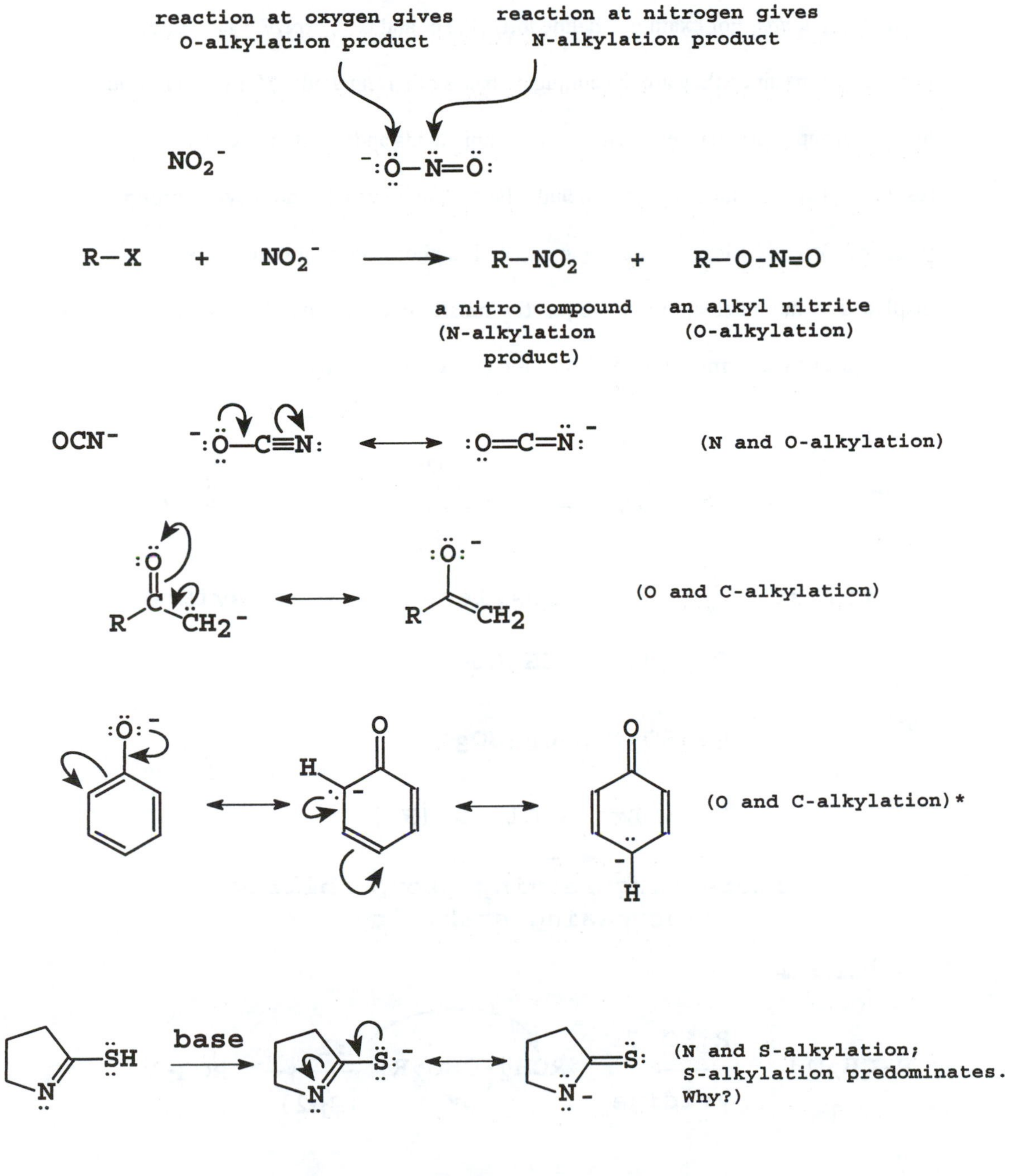

reaction at oxygen gives   reaction at nitrogen gives
O-alkylation product   N-alkylation product

NO_2^- $^-\!:\ddot{O}\!-\!\ddot{N}\!=\!\ddot{O}:$

$R\!-\!X$ + NO_2^- $\longrightarrow$ $R\!-\!NO_2$ + $R\!-\!O\!-\!N\!=\!O$

                      a nitro compound   an alkyl nitrite
                      (N-alkylation   (O-alkylation)
                      product)

OCN^- $^-\!:\ddot{O}\!-\!C\!\equiv\!N:$ $\longleftrightarrow$ $:O\!=\!C\!=\!\ddot{N}:^-$ (N and O-alkylation)

    $R\!-\!\overset{\displaystyle :\ddot{O}:}{\underset{\displaystyle CH_2^-}{C}}$ $\longleftrightarrow$ $R\!-\!\overset{\displaystyle :\ddot{O}:^-}{C}\!=\!CH_2$ (O and C-alkylation)

(O and C-alkylation)*

base   (N and S-alkylation;
                       S-alkylation predominates.
                       Why?)

*Note that C-alkylation leads to loss of aromaticity

91

(3) Nature of the Leaving Group (L)

The group that is displaced by a nucleophile (:Y⁻) in an S_N2 reaction is referred to as the *leaving group. Good leaving groups are stable anions (conjugate bases) derived from strong acids.* For example, halide ions (I⁻, Br⁻ and Cl⁻, *except* F⁻) are good leaving groups since they are the conjugate bases of strong acids. Many other good leaving groups are derived from strong organic acids and are stabilized by resonance (Minireview 2). These include the triflate, mesylate and tosylate groups (vide infra). The hydroxyl group of an alcohol (a poor leaving group that is not displaced by any nucleophile) is frequently transformed into one of these leaving groups prior to carrying out an S_N2 reaction (Scheme I below).

$$CF_3-\overset{\overset{\displaystyle O}{\|}}{\underset{\underset{\displaystyle O}{\|}}{S}}-O^- \quad > \quad CH_3-\!\!\!\left\langle\!\!\bigcirc\!\!\right\rangle\!\!\!-\overset{\overset{\displaystyle O}{\|}}{\underset{\underset{\displaystyle O}{\|}}{S}}-O^- \quad > \quad CH_3-\overset{\overset{\displaystyle O}{\|}}{\underset{\underset{\displaystyle O}{\|}}{S}}-O^-$$

$$\textbf{triflate} \qquad\qquad \textbf{tosylate} \qquad\qquad \textbf{mesylate}$$

$$CF_3COO^- \;>\; CH_3COO^-$$

$$CF_3SO_3^- \;>\; CH_3SO_3^-$$

$$I^- \;>\; Br^- \;>\; Cl^- \;>\; (F^-)$$

$$\longleftarrow$$

**increasing leaving group ability
(increasing stability)**

<u>**Scheme I**</u>

$$R\text{-}CH_2OH \quad\underset{\textbf{pyridine}}{\overset{R_1SO_2Cl}{\longrightarrow}}\quad RCH_2\text{-}OSO_2R_1 \quad\underset{(S_N2)}{\overset{:Y^-}{\longrightarrow}}\quad RCH_2Y$$

(4) Nature of the Solvent

The nature of the solvent used in carrying out an S_N2 reaction can have a dramatic effect on the rate and success of the reaction. S_N2 reactions are greatly facilitated by dipolar aprotic solvents such as dimethyl sulfoxide (DMSO), dimethyl formamide (DMF) and acetonitrile. These solvents enhance the nucleophilicity of anions by solvating the cation (solvation refers to the clustering of solvent molecules around the anion or cation).

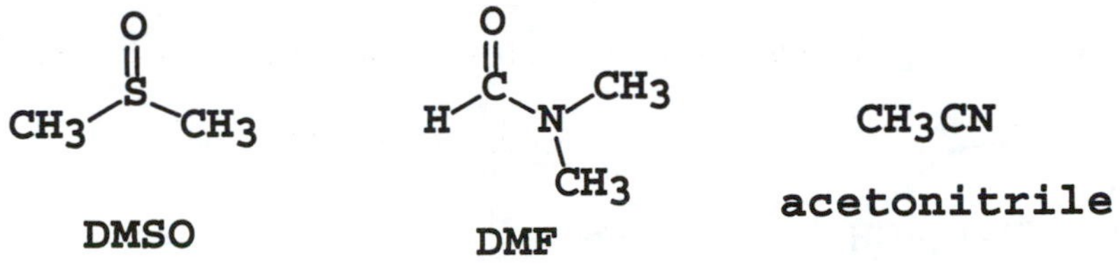

DMSO   DMF   acetonitrile

The nucleophilicity of an anion can also be enhanced by using a *crown ether* that complexes the cation, as shown below. The size of the cavity in each polyether determines the type of cation that is complexed. Inorganic reagents, such as potassium permanganate, which are ordinarily insoluble in organic solvents, become soluble in organic solvents in the presence of crown ethers.

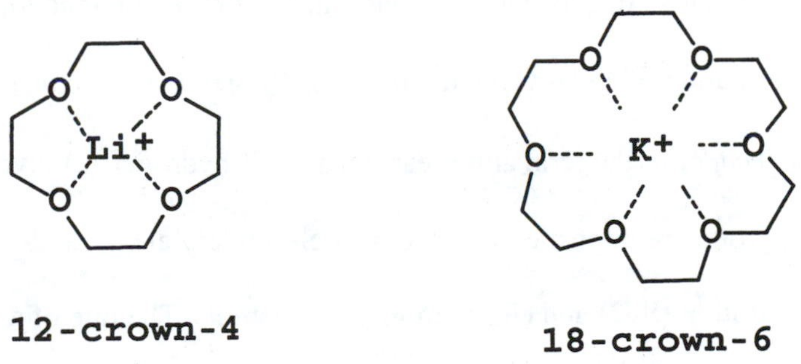

12-crown-4   18-crown-6

Note S$_N$2 reactions can be *intermolecular* (eq 1) or *intramolecular* (eq 2). Intramolecular S$_N$2 reactions take place within the same molecule. Reactions in which the nucleophile and reactive center are tethered together, are favored entropically and proceed at a much higher rate than the corresponding intermolecular reactions. Intramolecular reactions lead to the formation of cyclic structures, consequently, 5- and 6-membered ring structures are favored since they are free of ring strain.

$$Y:^- \quad + \quad R-L \quad \longrightarrow \quad R-Y \quad + \quad :L \qquad eq\ 1$$

$$^-:Y-(CH_2)_n-CH_2-L \quad \longrightarrow \quad \underset{(CH_2)_n}{\overset{Y-CH_2}{|\,/}} \quad + \quad :L \qquad eq\ 2$$

B. Elimination (E$_2$) Reactions

(1) As mentioned earlier, when an alkyl halide or similar type of compound (R-L, where L = halogen or other good leaving group) is treated with a strong base (:Y$^-$), an S$_N$2 reaction takes place. This is the course that an S$_N$2 reaction typically follows when a methyl, allylic, or benzylic halide is used. However, when a 3° R-L is used, the reaction follows a different course due to *steric* reasons. An E$_2$ elimination reaction takes place, leading to the formation of a C = C bond (eq 3). *An E$_2$ reaction is a concerted 1,2- or β- anti elimination reaction* (the term anti means that the β-hydrogen removed by the base and the leaving group are opposite to each other). Secondary alkyl halides undergo competing substitution (S$_N$2) and elimination (E$_2$) reactions. The rate of an E$_2$ reaction increases with increasing base strength, leaving group ability, and the relative stability of

the product(s). Typical bases used in E_2 reactions include HO^-, RO^-, and bulky organic

bases such as diisopropyl ethyl amine, DBU and DBN.

$$Y:^- \quad \curvearrowright \qquad \underset{L}{\overset{H}{\underset{\displaystyle C-C}{}}} \quad \longrightarrow \quad C=C \quad + \quad BH \quad + \quad :L \qquad Eq \ 3$$

$$(L = halogen, \ OSO_2R, \ etc.)$$

(2) Thermal Syn Elimination Reactions

Thermal *syn elimination* reactions constitute another somewhat less common type of β-

elimination reaction. This type of reaction proceeds thermally and through a cyclic

transition state, and does not require the use of base (the term syn means that the β-

hydrogen and the leaving group are on the same side) . Syn eliminations involving

sulfoxides and selenoxides are widely used in the facile formation of C=C double

bonds. While other leaving groups have been used in the past (acetate, amine oxide, etc.),

elimination occurs at much higher temperatures.

$$\underset{\displaystyle C-C}{\overset{\displaystyle H \quad \overset{O}{\underset{\parallel}{}} \quad X}{}} \quad \longrightarrow \quad C=C \quad + \quad HOX$$

$$(X = SR \ or \ SeR)$$

95

<u>MINIREVIEW 6</u> Chemical Reactivity and Ring Strain

Three and four-membered rings have considerable *ring strain*, namely, the ring bonds are weak because of poor overlap of the atomic orbitals and, consequently, exhibit high reactivity. They are highly reactive toward Lewis bases (nucleophiles), giving rise to ring-opened products. Note that *nucleophilic ring opening of an epoxide, for example, involves back-side attack at the least substituted (least congested) carbon, and proceeds with inversion of configuration* (eq 1). Transient 3-membered species, such as halonium and selinarium ions, behave similarly (eq 2 and 3).

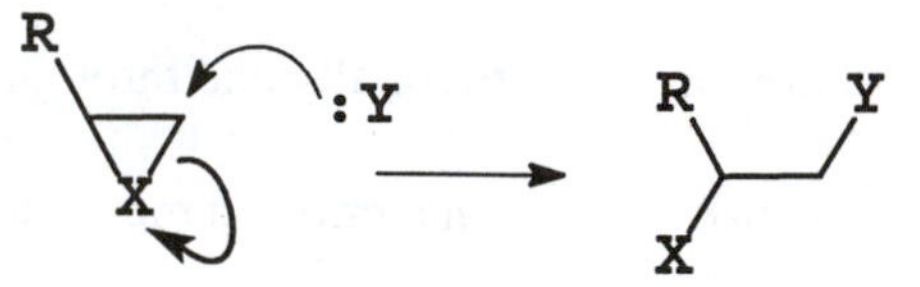

eq 1

(X = O (epoxide), S (episulfide), and NH (aziridine)

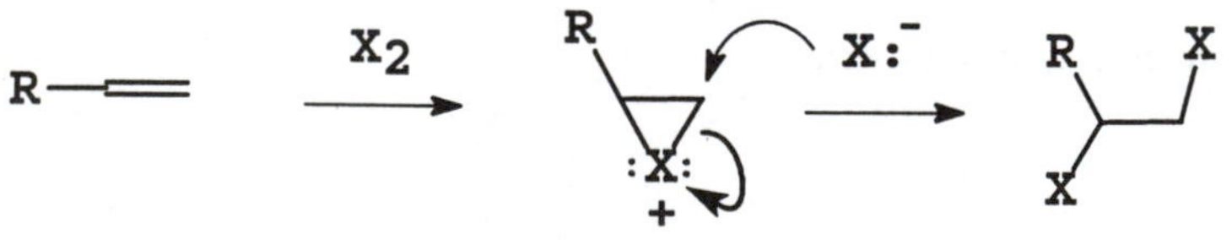

eq 2

(X_2 = Cl_2, Br_2, I_2)

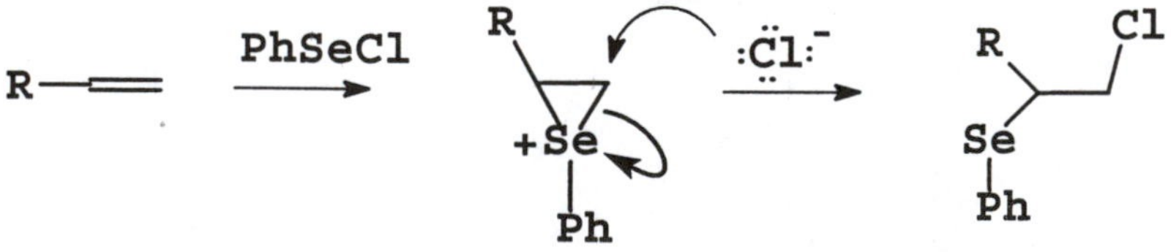

eq 3

96

3- and 4-membered cyclic esters (lactones) and cyclic amides (lactams) also react rapidly with nucleophiles (eq 4).

$$(X = O, NH, NR)$$

Example 1

The highly reactive β-lactam ring of penicillins and related antibacterial agents undergoes ring opening readily. Hydrolysis of the β-lactam ring in penicillins results in total loss of antibacterial activity (in practical terms, suspensions of penicillins, such as pivampicillin, are kept refrigerated in order to slow down the hydrolysis reaction. They are thrown away after a week or so since they contain mostly inactive penicillin).

inactive penicillin

Example 2

The toxicity associated with a drug or substance is frequently due to a highly reactive metabolite. For example, the toxicity and carcinogenicity of polycyclic aromatic hydrocarbons and other aromatic compounds, is attributed to the formation of highly

reactive epoxides by liver-metabolizing enzymes, and their subsequent reaction with

nucleophilic cellular components, including DNA.

QUESTIONS 35-50

Questions 35-50 are meant to help you gain a better understanding of

*Generation of anions/Bronsted acidity;

*S_N2 reactions (mechanism, nucleophilicity, ambident nucleophiles, leaving groups, stereochemistry, solvent effects, etc.);

*E_2 reactions (anti and thermal syn elimination reactions);

*Nucleophilic ring-opening reactions.

35)

36)

37)

38)

100

39)

40)

41)

42)

43)

44)

45)

46)

47)

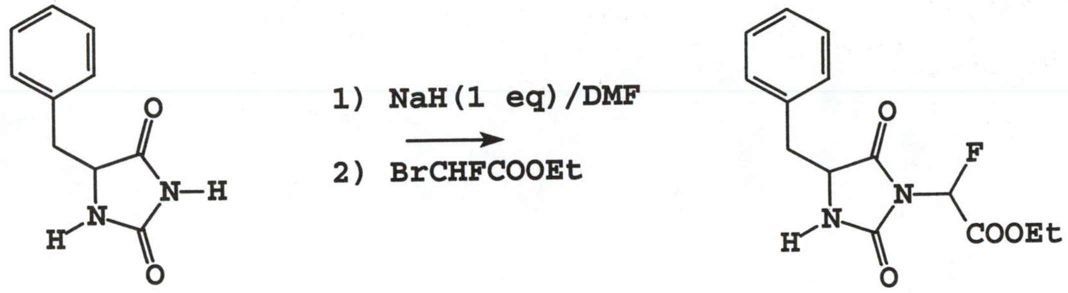

O
N—H
H
N
O
1) NaH(1 eq)/DMF
2) BrCHFCOOEt
O
F
N
H
N
COOEt
O

48)

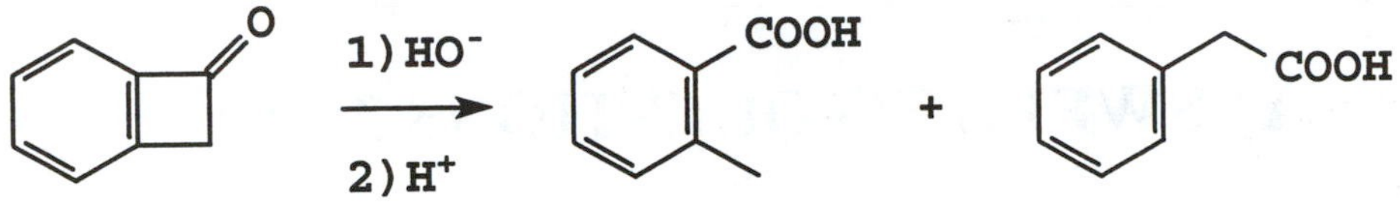

O
1) HO⁻
2) H⁺
COOH
+
COOH

49)

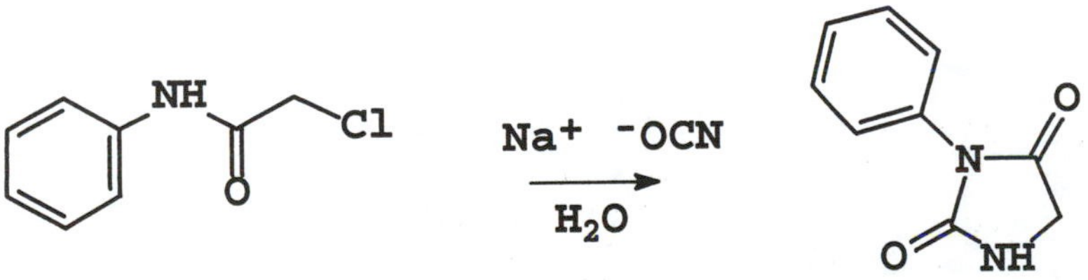

NH
Cl
O
Na+ ⁻OCN
H2O
O
N
O
NH

50)

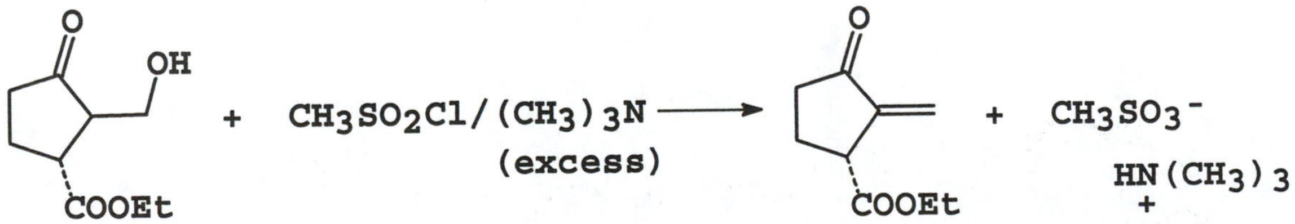

O
OH
COOEt
+
CH3SO2Cl/(CH3)3N
(excess)
O
COOEt
+
CH3SO3⁻
HN(CH3)3
+

ANSWERS TO QUESTIONS 35-50

35)

An initial fast acid-base reaction yields the anion
(LB or nucleophile). This is followed by an intramolecular
S_N2 reaction. The reaction proceeds readily at room
temperature despite the fact that the carboxylate anion
is a weak nucleophile because (a) the reaction is
intramolecular (reaction takes place within the same
molecule) and therefore favored entropically. Furthermore,
the reacting groups are arranged in a favorable geometrical
alignment; and, (b) the reaction involves a highly reactive
allyl bromide (Minireview 5).

36)

Formation of the anion is followed by an intramolecular S_N2
reaction. In general, intramolecular ring formation
(cyclization) is favored when the reaction is carried out
in very high dilution. The relative ease of ring formation
is dependent on many factors, including the stability of
the ring to be formed, with lowest yields obtained with
medium size (8 to 11-membered) rings.

37)

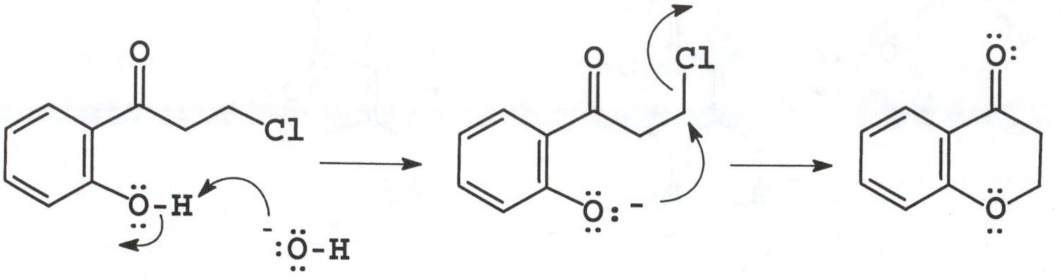

Formation of the anion of phenol (pK_a ~10) is rapid and quantitative. This is followed by an intramolecular S_N2 reaction. Note that the hydroxide ion could, in principle, displace the chloride first. It should be remembered, however, that the rate of organic reactions follows the order

Bronsted acid/base > S_N2 ~ nucleophilic addition > nucleophilic

    reactions   acyl reactions

                   (one step)   (two steps)

38)

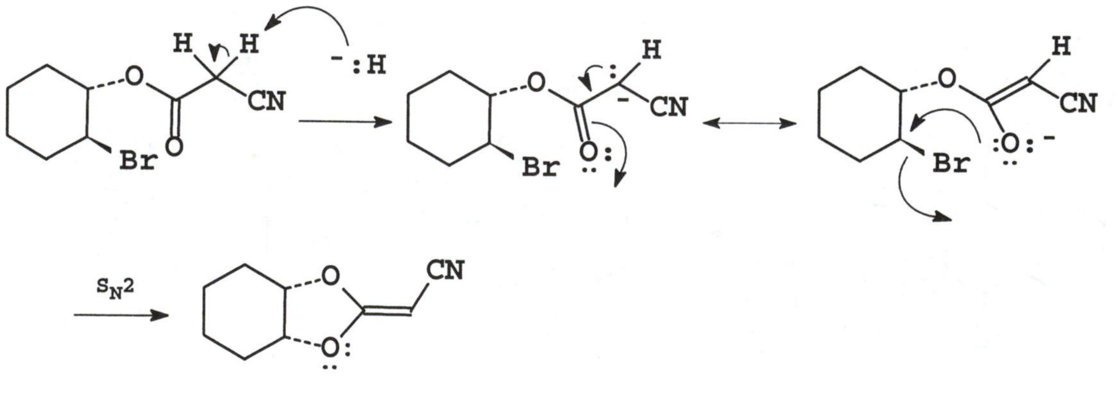

Rapid and quantitative formation of the anion using sodium hydride is followed by an intramolecular S_N2 reaction. As is true for all S_N2 reactions, the reaction proceeds with inversion of configuration, accounting for the observed stereochemistry. See Minireview 4 for a discussion of acidity and anion formation.

39)

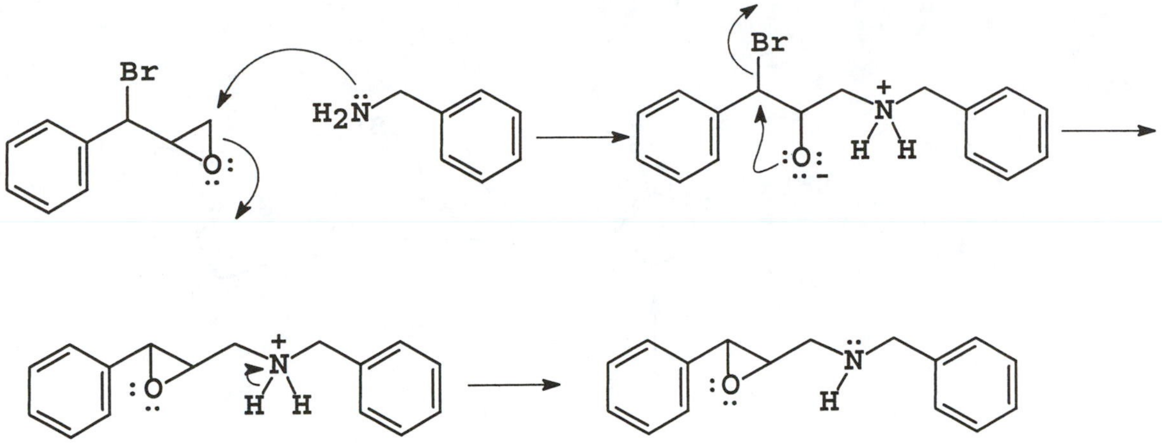

Epoxides (oxiranes) are highly reactive (ring strain) and, consequently, undergo ring opening reactions with the nucleophile attacking the least substituted (least congested) carbon.

40)

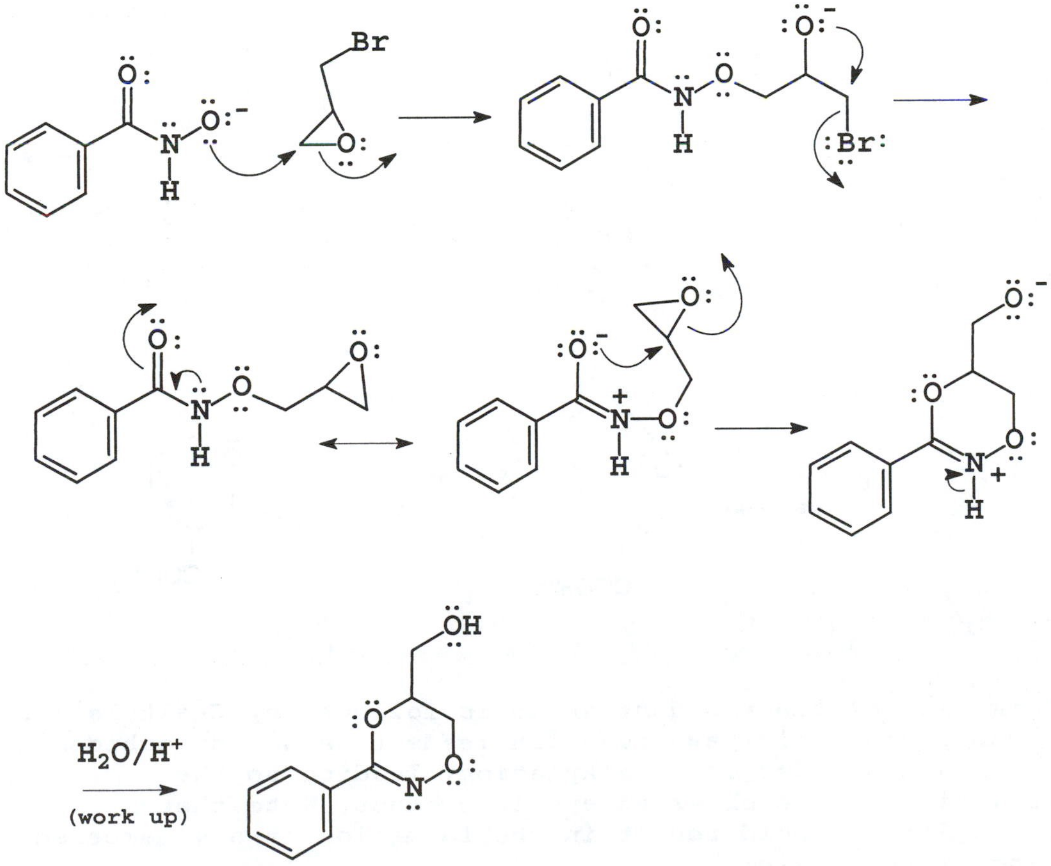

$$\xrightarrow[\text{(work up)}]{H_2O/H^+}$$

41)

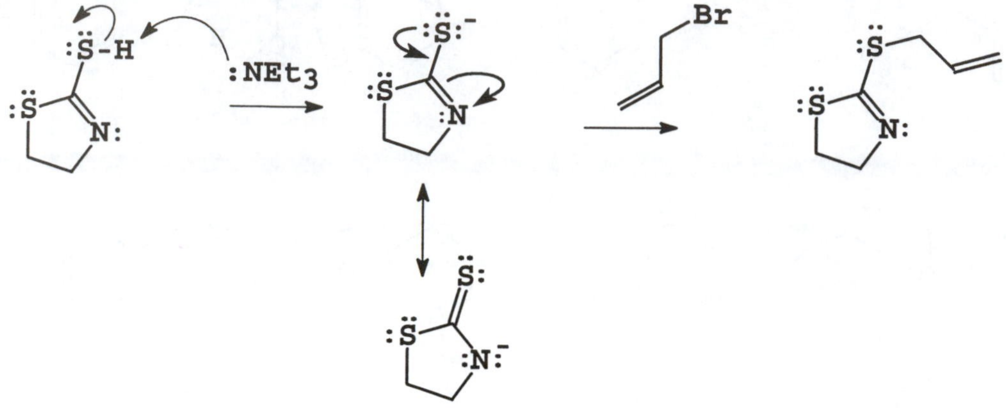

The anion generated in the first step (see Minireview 4) is an
ambident anion, i.e., it can react at either the N or S atom.
The S_N2 reaction takes place on S, attesting to the greater
nucleophilicity of sulfur (see Minireview 5).

42)

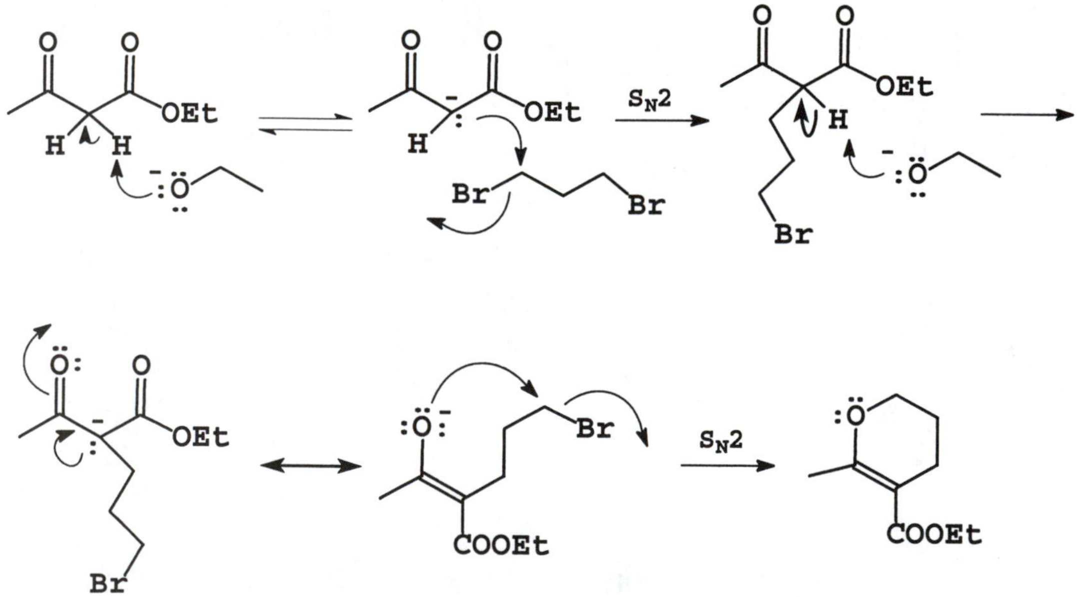

Formation of the ambident anion is followed by C-alkylation.
A subsequent acid/base reaction leads to a second ambident
anion which undergoes O-alkylation, leading to the
formation of the observed cyclic product. Note that
C-alkylation would result in the formation of a 4-membered
ring (ring strain).

43)

The first step involves a rapid and quantitative formation of a resonance-stabilized ambident anion. A subsequent intramolecular S_N2 reaction (C-alkylation) leads to the observed tricyclic non-aromatic product. As pointed out earlier, the nature of the product formed in reactions involving ambident anions is dependent on a range of factors.

44)

45)

intramolecular S_N2

110

46)

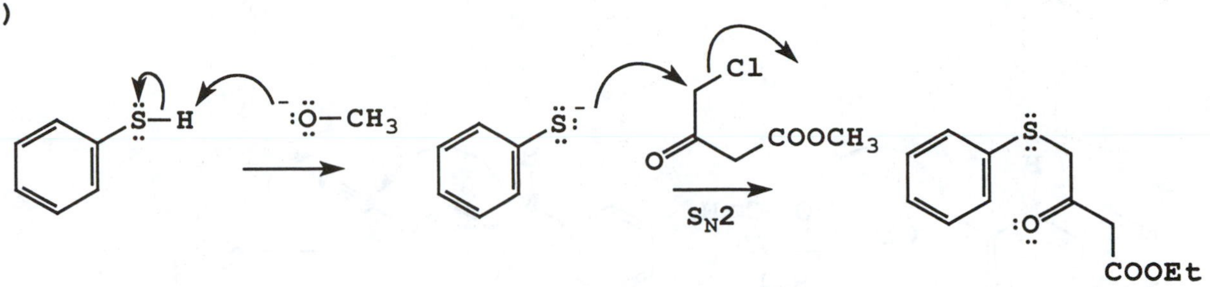

47)

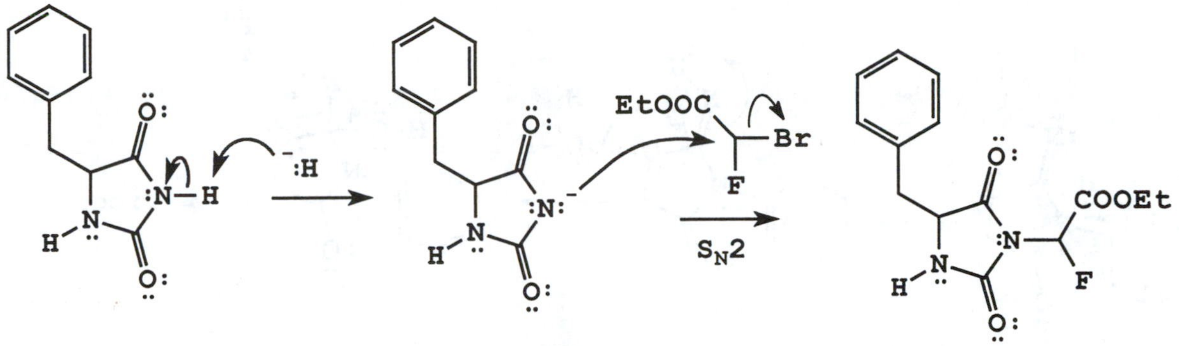

Why does the base remove the hydrogen shown and not the other one?
Why does the anion displace the bromine (as Br⁻) instead of the
fluorine (as F⁻)?

48)

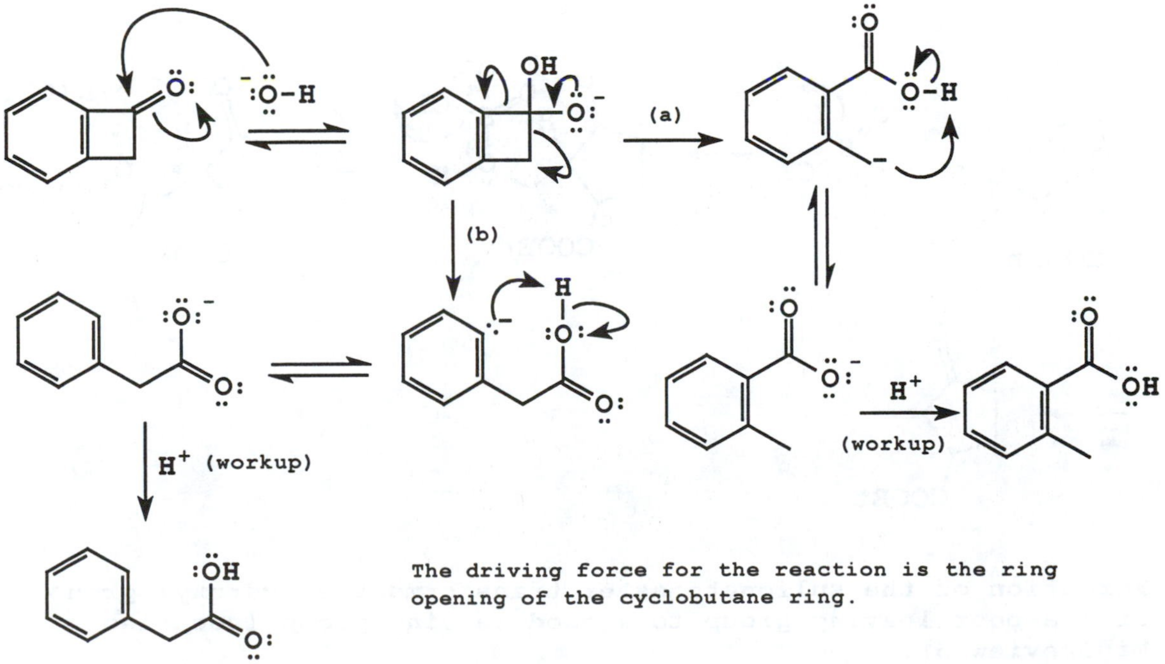

The driving force for the reaction is the ring
opening of the cyclobutane ring.

111

49)

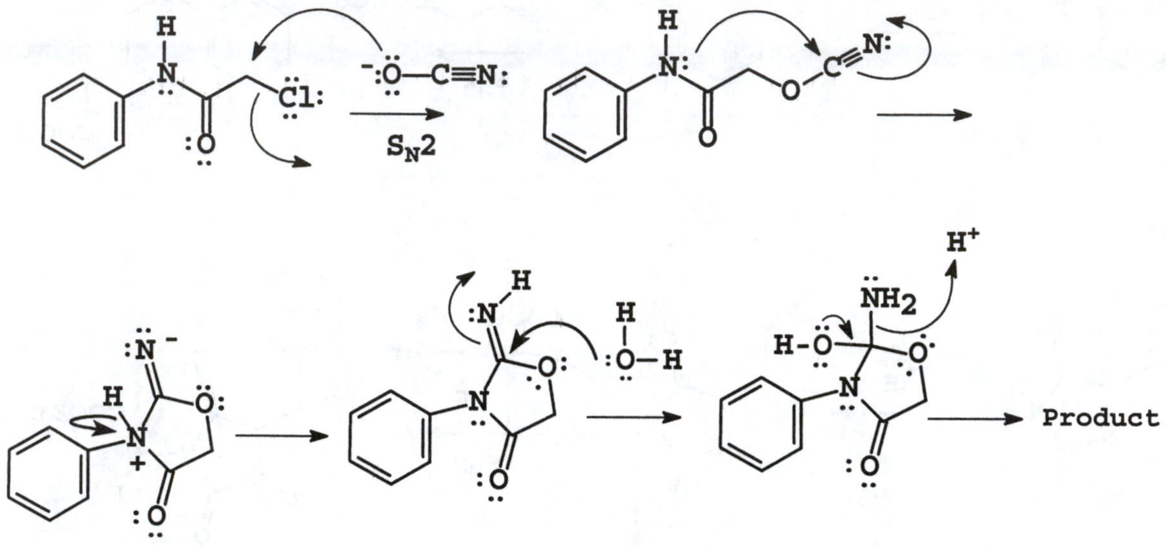

50)

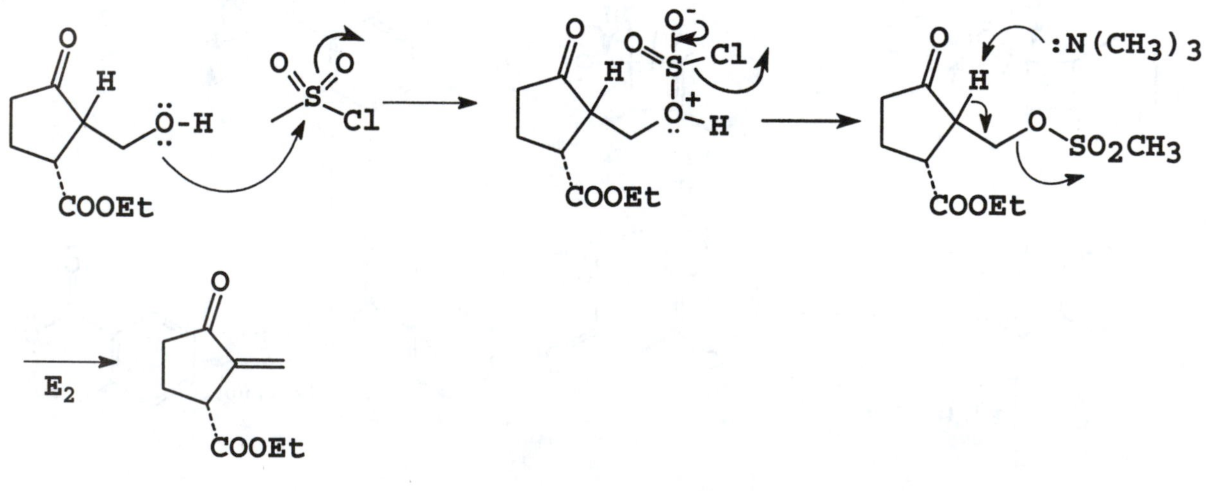

Formation of the sulfonate ester transforms the hydroxyl group from a poor leaving group to a good leaving group (see also Minireview 5).

Minireview 7 focuses on nucleophilic addition reactions (the typical reaction of aldehydes and ketones). This should be studied prior to attempting questions 51-70.

<u>MINIREVIEW 7</u> Nucleophilic Addition Reactions of Aldehydes and Ketones

The typical reaction of aldehydes and ketones is *nucleophilic addition*. Notice that a nucleophilic addition reaction is essentially a Lewis acid/Lewis base reaction involving a Lewis base (:Y$^-$ or :Y) with the electron-deficient carbon (Lewis acid) of the carbonyl group. The addition product arising from strong nucleophiles is then protonated (typically during the workup of the reaction) and the final product isolated (eq 1). Typical strong nucleophiles include Grignard reagents (R MgX or R:$^-$ Mg^{2+}X$^-$), alkyl lithium reagents (RLi or R:$^-$ Li$^+$), cyanide ion, alkyne anions (RC≡C:$^-$), reducing agents that deliver hydride ($^-$:H) ions, etc.

eq 1

In the familiar *aldol condensation reaction*, an aldehyde or ketone having an acidic α-hydrogen reacts with dilute base to yield a carbon anion (Lewis base), which then adds to the carbonyl carbon of a second molecule of the aldehyde or ketone, leading to the formation of a β-hydroxy aldehyde or ketone (an α,β-unsaturated aldehyde or ketone may also form depending on the reaction conditions) (Scheme I).

In the case of some nucleophilic addition reactions, the initial nucleophilic addition product reacts further to yield a more stable product. For example, the reaction of an

114

Scheme I

R-CH$_2$-C(=O)-H(R) ⇌ (HO$^-$) R-CH$^-$-C(=O)-H(R) (Y:$^-$) → (R)H-C(=O)-CH$_2$R ⇌ RCH$_2$-C(O$^-$)(H(R))-CH(R)-C(=O)-H(R)

⇌ RCH$_2$-C(OH)(H(R))-CH(R)-C(=O)-H(R) → RCH$_2$-CH=C(R)(H(R))-C(=O)-H(R)

aldehyde or ketone with a Wittig reagent involves an initial nucleophilic addition reaction. An intramolecular Lewis acid/base reaction yields a 4-membered intermediate which collapses, forming an alkene (Scheme II).

Scheme II

R-CH$_2$-C(=O)-H(R) + $^-$:C(R)(R)-$^+$PPh$_3$ → R-CH$_2$-C(O$^-$)(H(R))-C(R)(R)-$^+$PPh$_3$ → R-CH$_2$-C(O)(H(R))-C(R)(R)-PPh$_3$

→ RCH$_2$-CH=C(R)(R)(H(R)) + O=PPh$_3$

Likewise, *primary* amines react readily with aldehydes and ketones, however, the initial adduct loses a molecule of water to form an *imine* (also called a *Schiff's base*) (Scheme III). Imines undergo addition reactions with Lewis bases. These reactions proceed readily when the imine is protonated (Scheme IV). *Secondary* amines react with aldehydes and ketones to form *enamines* (Scheme V).

Scheme III

$$\overset{R}{\underset{(R)H}{}}C=O \quad :NH_2-R_1 \longrightarrow \overset{R}{(R)H-\underset{\underset{R_1}{\overset{+}{N}}-H}{\overset{|}{C}}-O:^-} \rightleftharpoons \overset{R}{(R)H-\underset{\underset{R_1}{\overset{}{N}-H}}{\overset{|}{C}}-OH} \xrightarrow{H^+} \overset{R}{\underset{(R)H}{}}C=N-R_1 \; + \; H_2O$$

(prototropic shift)

imine

Scheme IV

$$\overset{R}{\underset{(R)H}{}}C=\overset{..}{N}-H \underset{(pH<7)}{\overset{H^+}{\rightleftharpoons}} \overset{R}{\underset{(R)H}{}}C=\overset{+}{\underset{H}{N}}-H \xrightarrow{:Y} \overset{Y}{R-\underset{\underset{(R)H}{}}{\overset{|}{C}}-\overset{..}{N}H_2}$$

$$\overset{R}{\underset{(R)H}{}}\overset{+}{C}-\overset{H}{\underset{H}{\overset{}{N}}}:$$

Scheme V

$$R-CH_2-\overset{O}{\underset{H}{\overset{\|}{C}}} \; \overset{HNR_1R_2}{\rightleftharpoons} \; R-CH_2-\overset{OH}{\underset{H}{\overset{|}{C}}}-N\overset{R_1}{\underset{R_2}{}} \rightleftharpoons R-CH=CH-N\overset{R_1}{\underset{R_2}{}} \; + \; H_2O$$

enamine

The reaction of aldehydes and ketones with weak nucleophiles (:Y), such as alcohols and water, requires acid catalysis (Scheme VI). Recall that the reaction of two mols of an alcohol with an aldehyde or ketone in the presence of acid (dry HCl) leads to the formation of an acetal or ketal, respectively (Scheme VII).

Scheme VI

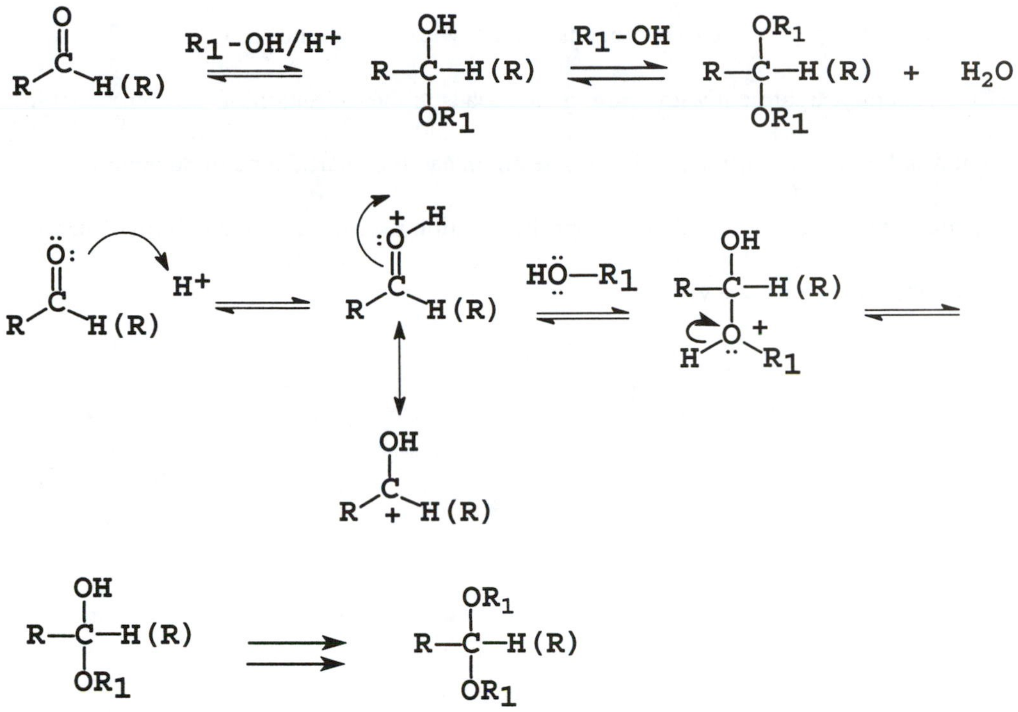

α,β-Unsaturated aldehydes and ketones undergo the Michael addition reaction with Lewis bases (nucleophiles) (Scheme VII). Related conjugated systems, R-CH=CH-X, where X = COOR, CN, SO$_2$R and NO$_2$, behave similarly.

Scheme VII

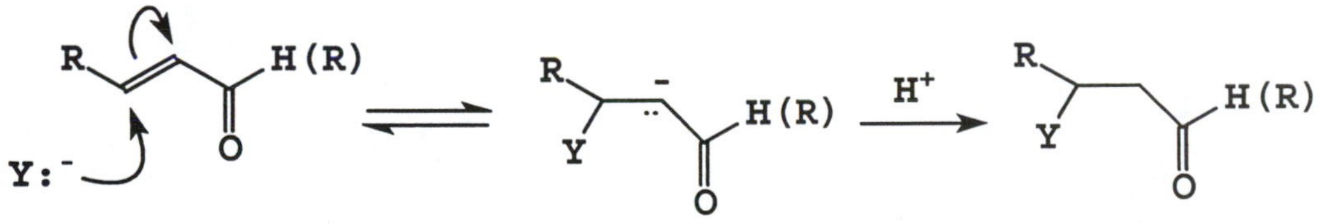

117

Lastly, the pK_a of the α-hydrogens in aldehydes and ketones is about 20.

Consequently, treatment with a strong base leads to the formation of the corresponding

anion (a Lewis base or nucleophile). The anion can then participate in nucleophilic

displacement reactions (S_N2), nucleophilic addition reactions, or nucleophilic substitution

reactions (see Minireview 4).

QUESTIONS 51-70

Questions 51-70 are meant to help you gain a better understanding of

*Acid and base-catalyzed addition reactions;

*Aldol condensation reactions;

*Michael addition reactions;

*Keto-enol tautomeric equilibria. _

51)

52)

53)

54)

55)

56)

57)

58)

59)

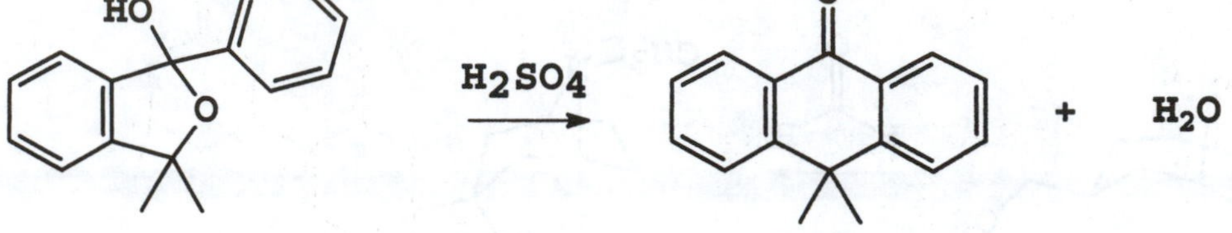

H_2SO_4 → + H_2O

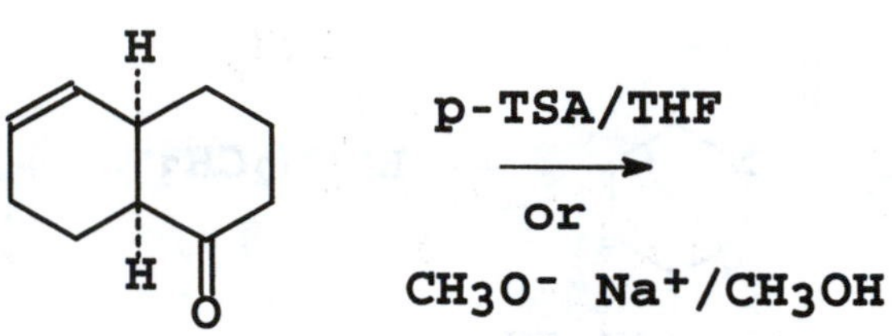

60)

$$\xrightarrow[CH_3O^- \ Na^+/CH_3OH]{p\text{-}TSA/THF \ \text{or}}$$

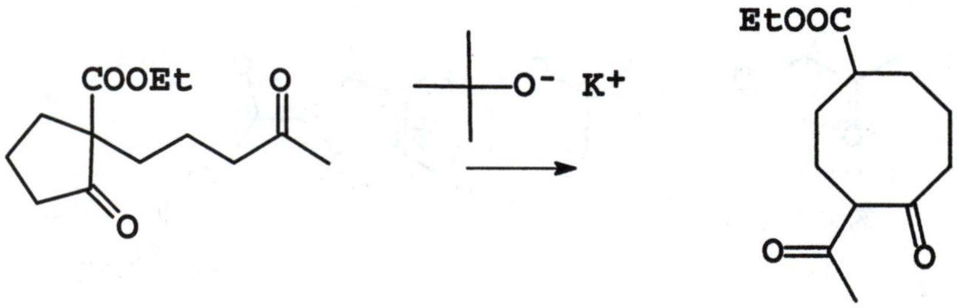

61)

62)

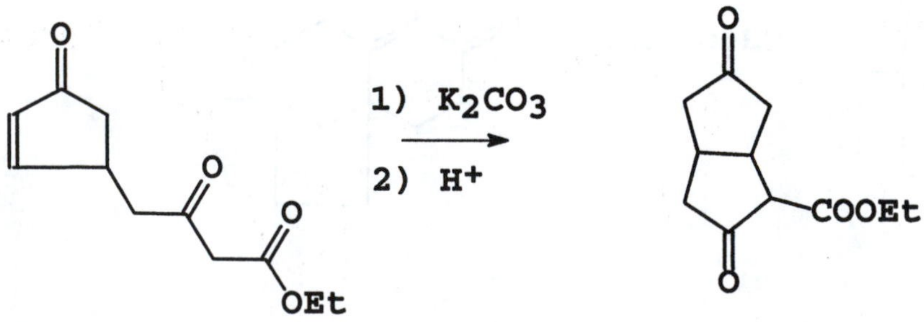

1) K_2CO_3
2) H^+

122

63)

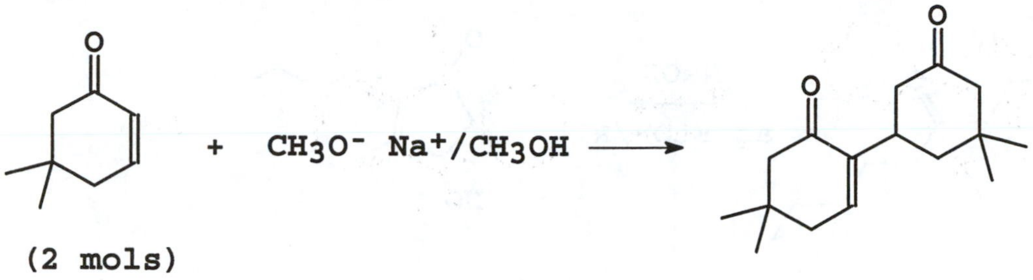

+ CH₃O⁻ Na⁺/CH₃OH ⟶

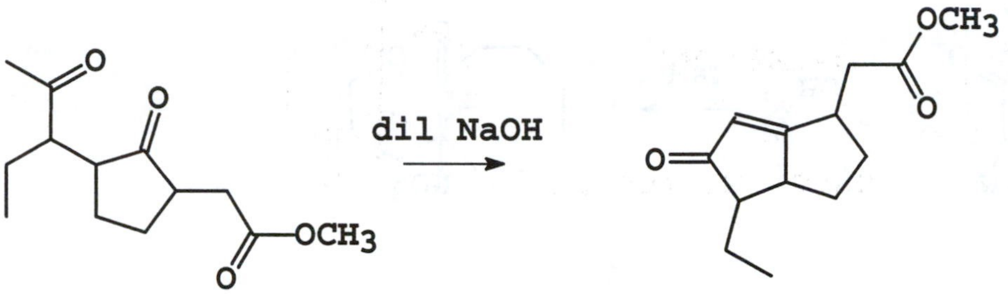

(2 mols)

64)

dil NaOH

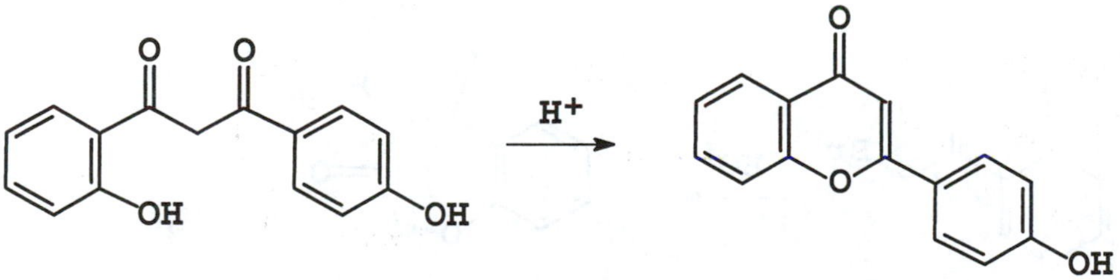

65)

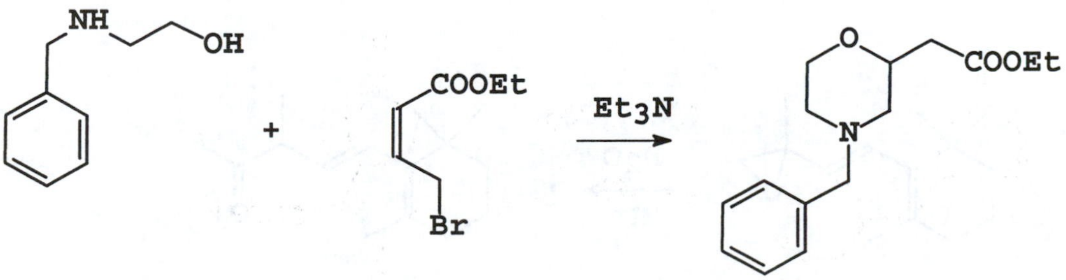

H⁺

66)

+ Et₃N

123

67)

68)

69)

70)

ANSWERS TO QUESTIONS 51-70

51)

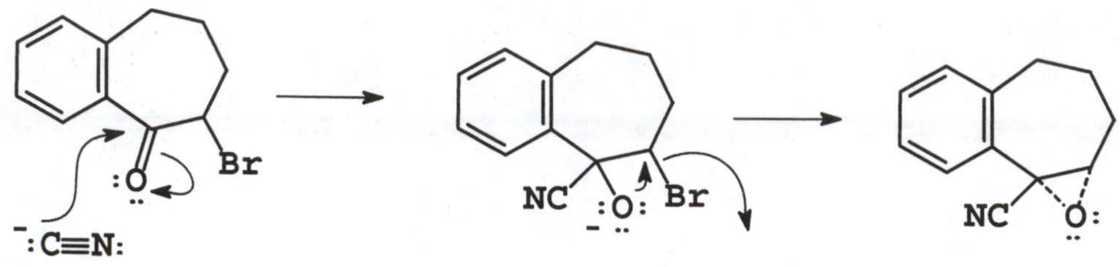

A straightforward nucleophilic addition is followed by an intramolecular S_N2 reaction.

52)

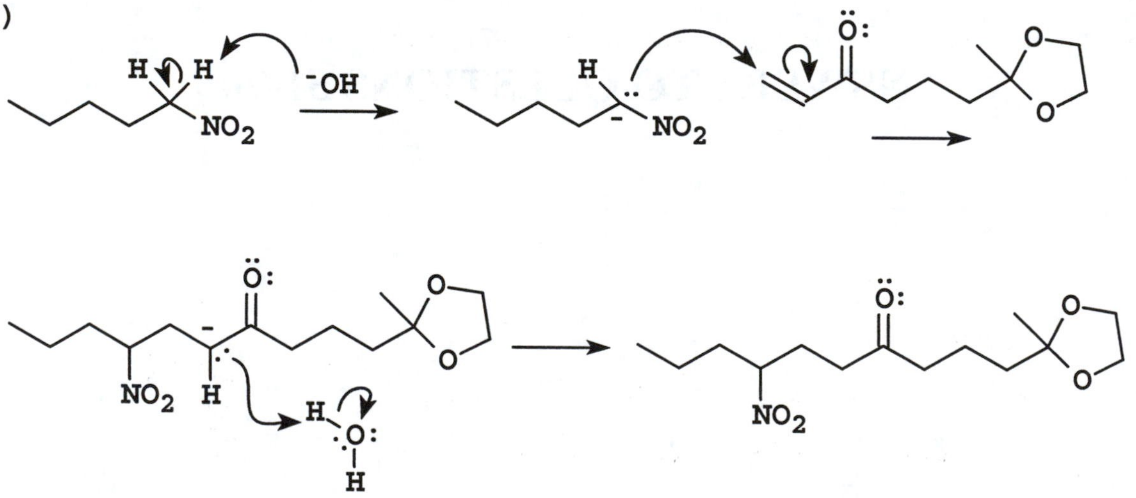

Note that once the anion is generated, what happens next is determined by the nature of the reactant (see Minireview 4). Since the reactant in this case is a conjugated carbonyl compound, a Michael addition reaction takes place.

53)

(intramolecular aldol condensation reaction)

54)

An S$_N$2 reaction is followed by an intramolecular Michael addition reaction.

55)

56)

128

57)

58)

129

59)

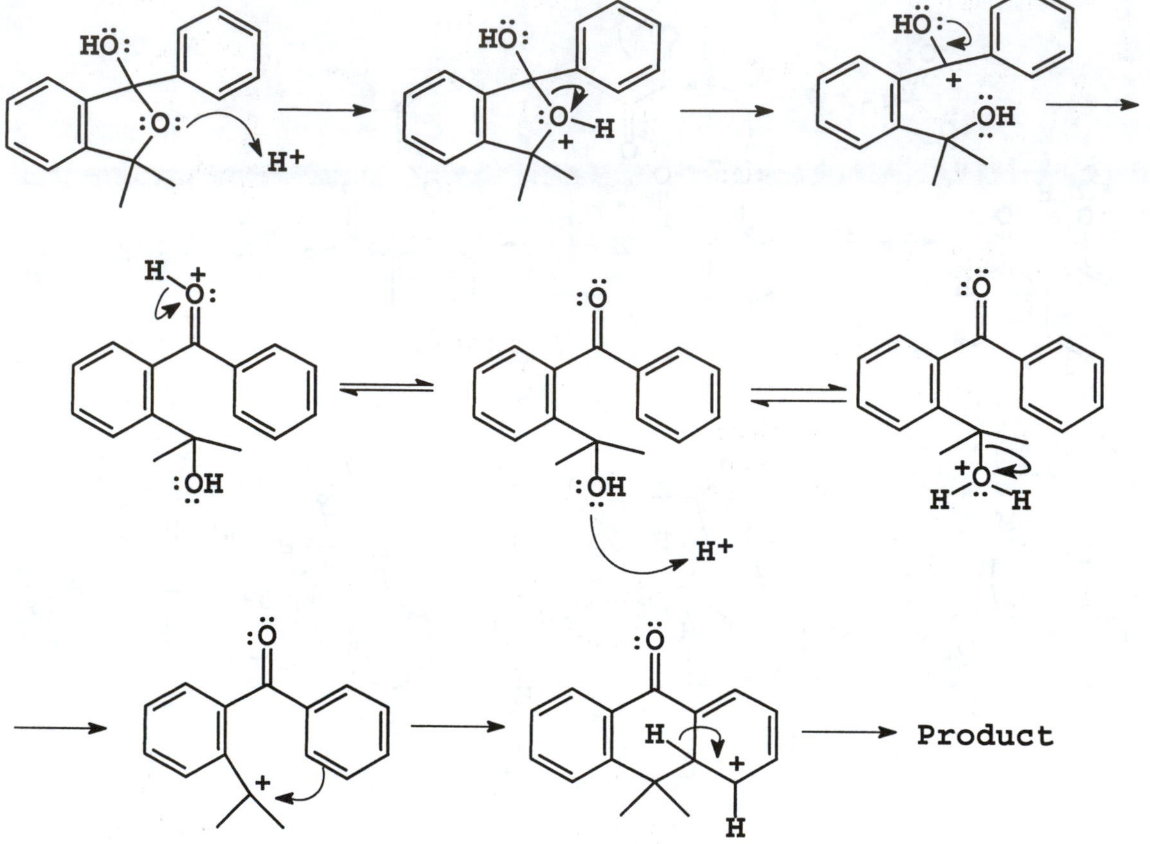

60)

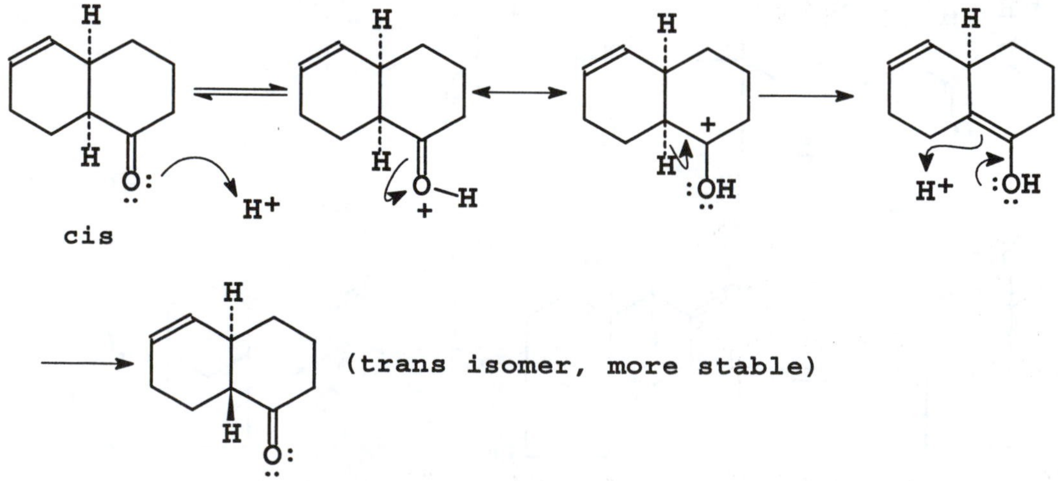

cis

(trans isomer, more stable)

61)

62)

Formation of the anion is followed by an
intramolecular Michael addition reaction

63)

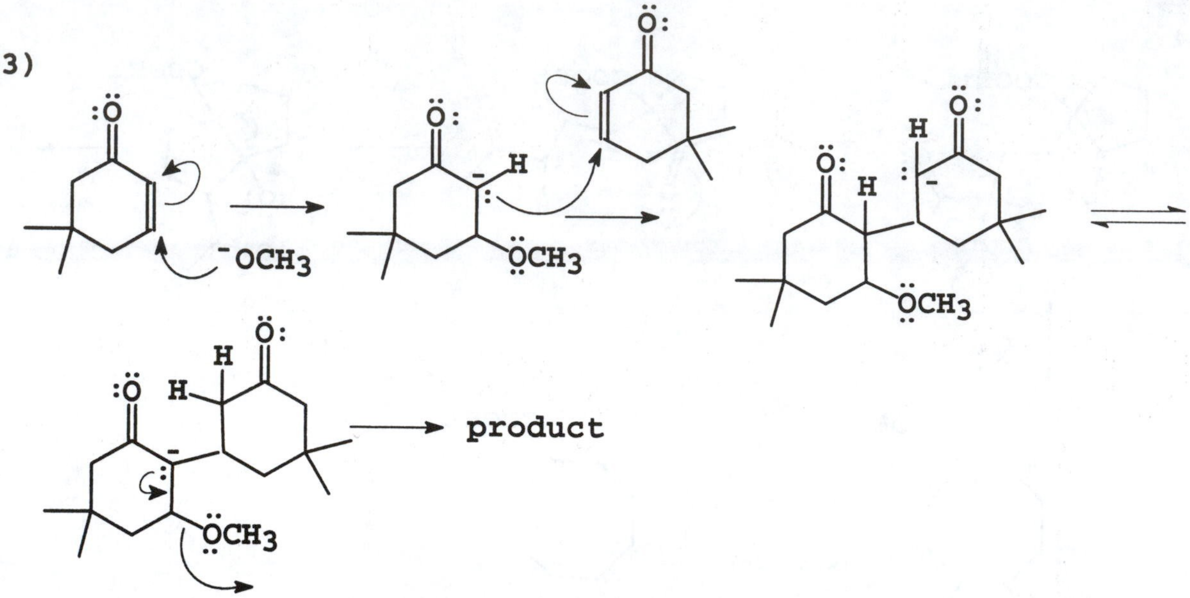

Two consecutive Michael addition reactions are followed by a prototropic shift (an acid-base reaction) and elimination of the methoxide ion.

64)

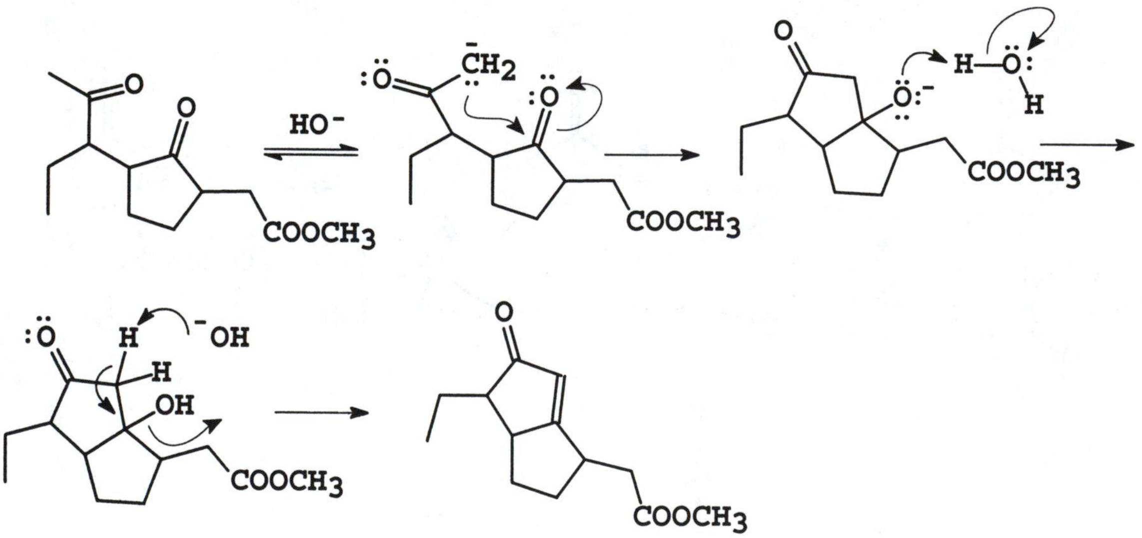

65)

66)

133

67)

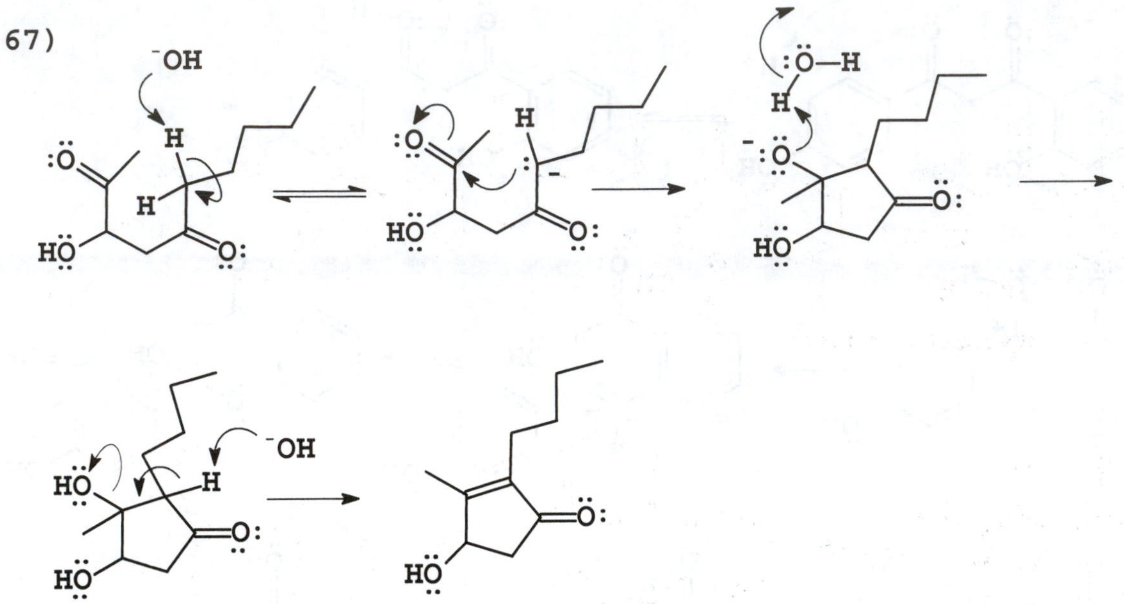

68)

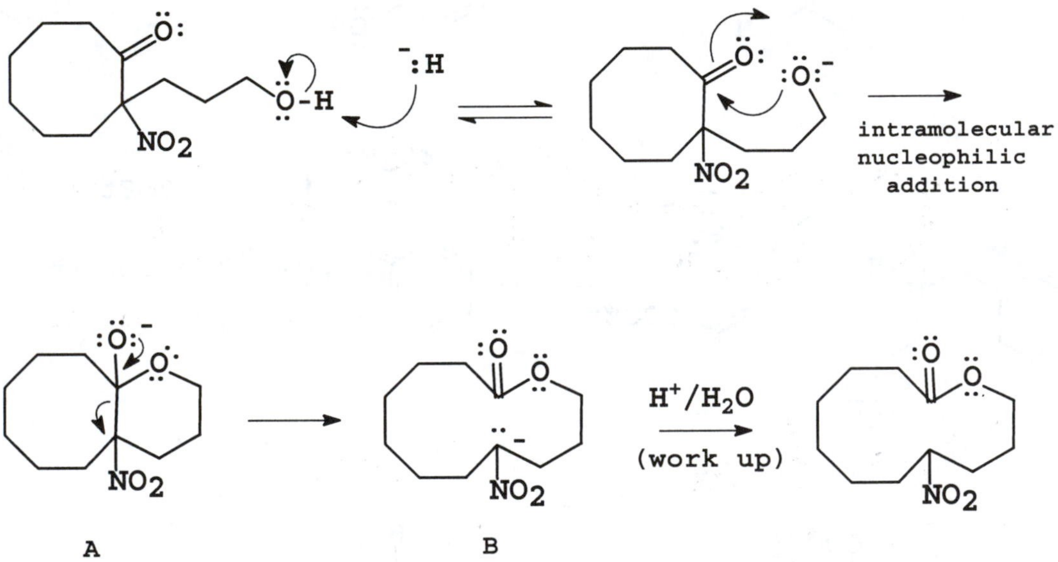

Species A is the expected nucleophilic addition product, however, instead of becoming protonated, A reforms the strong carbonyl bond, yielding a highly stable anion B.

Exercise Write all the resonance structures that can be written for anion B.

134

69)

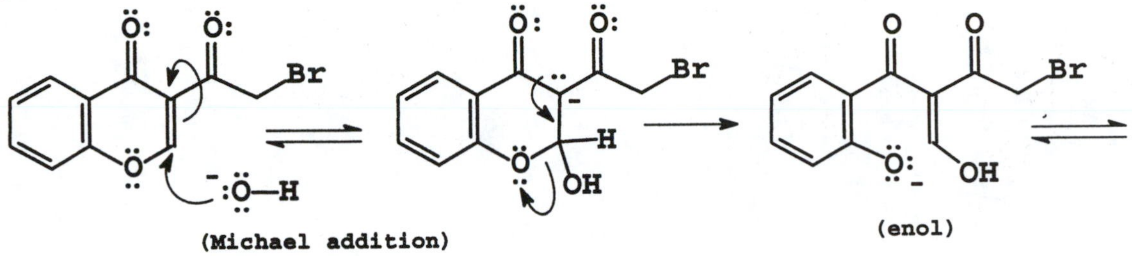

(Michael addition)

(enol)

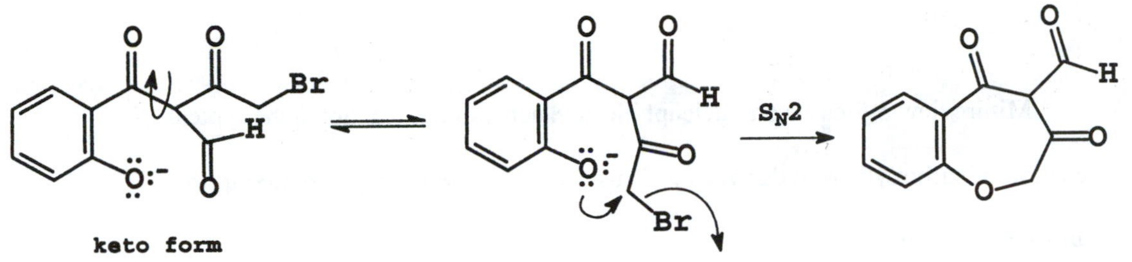

keto form

S_N2

70)

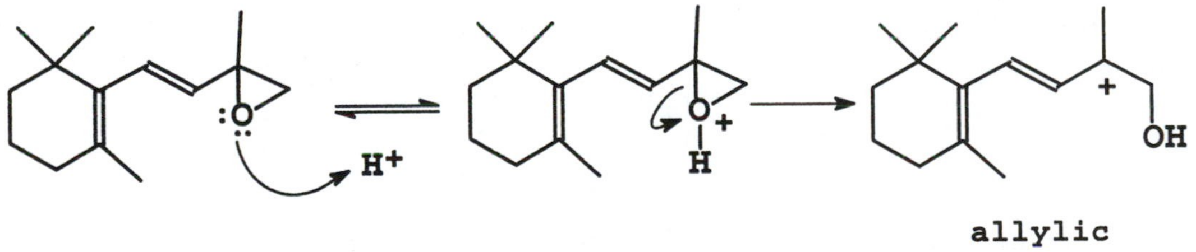

allylic

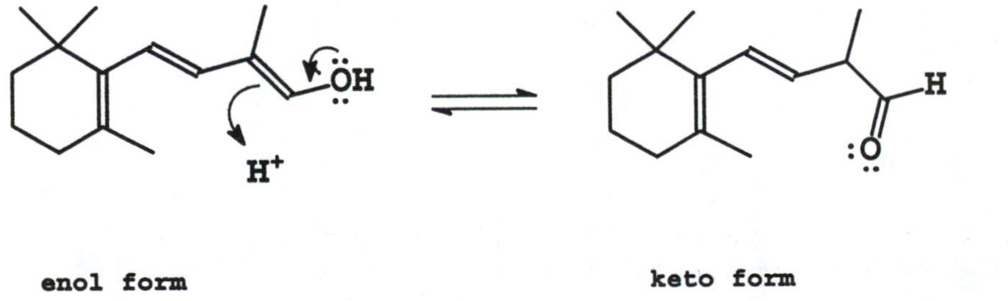

enol form

keto form

Minireview 8 focuses on nucleophilic acyl substitution reactions, the typical reaction of carboxylic acid derivatives. This should be studied prior to attempting questions 71-100.

<u>MINIREVIEW 8</u> Nucleophilic Acyl Substitution Reactions

The typical reaction of carboxylic acid derivatives (acid halides, anhydrides, esters, thioesters, and amides) with Lewis bases is *nucleophilic acyl substitution* (eq 1). The mechanism of this reaction is shown in Scheme 1 below.

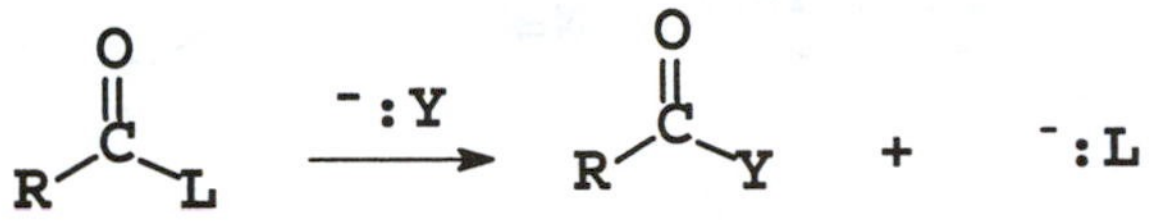

Mechanism

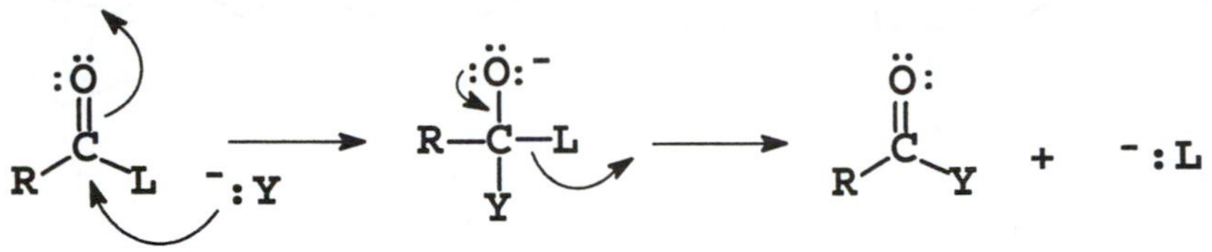

Reactions involving weak nucleophiles (:Y) can be speeded up by using acid catalysis (eq 2). With the exception of amines and phosphorus nucleophiles, nucleophiles that do not bear a negative charge (:Y) are classified as being weak, while those having a negative charge are classified as strong nucleophiles.

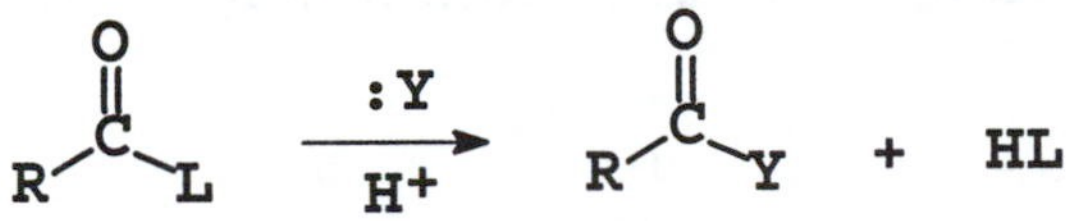

Mechanism

137

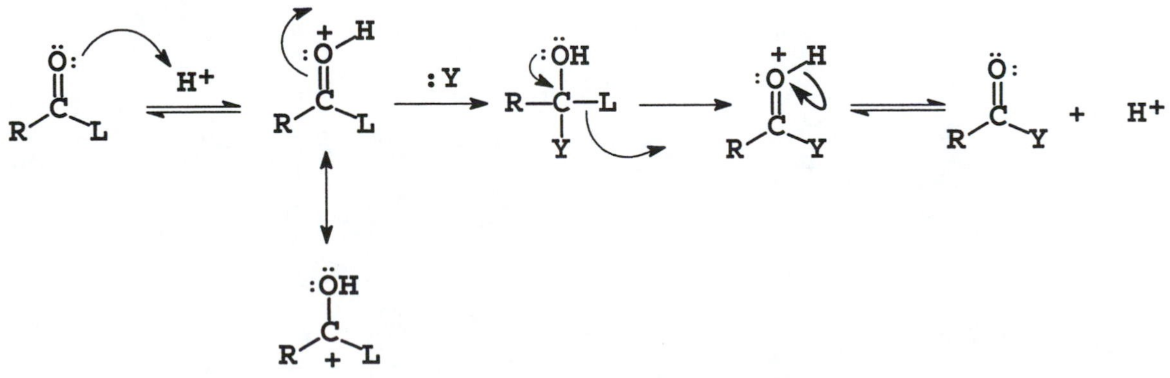

Exercises Using the general mechanistic principles shown above, write a **mechanism** for each of the following reactions:

1)

a)

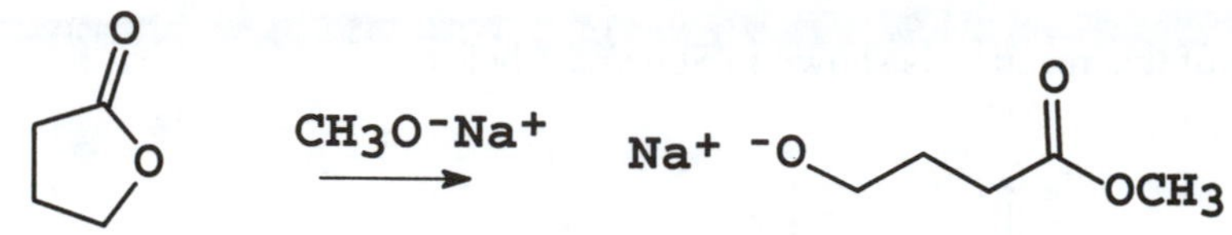

CH₃O⁻Na⁺

Na⁺ ⁻O ⟋⟍⟋ C(=O) OCH₃

b)

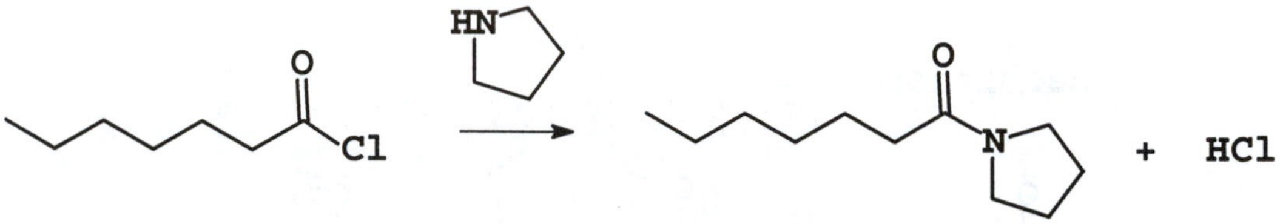

HN⟨ ⟩

→

+   HCl

c)

H₂O/H⁺

→

with COOH, COOH

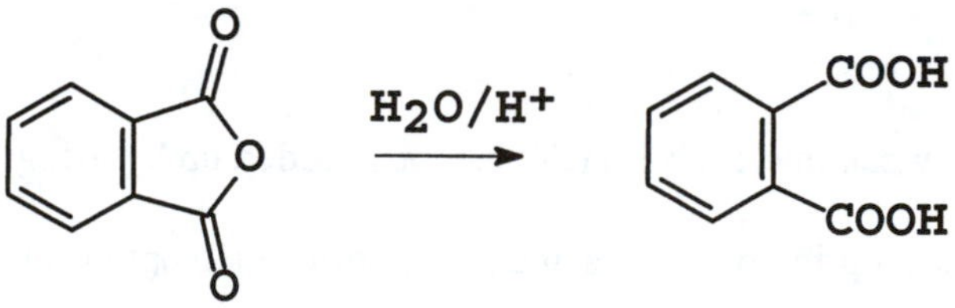

2) Predict the product of the following reaction:

NH₂ , CH₃ on ring

+

Br ⟋ C(=O) Br

→

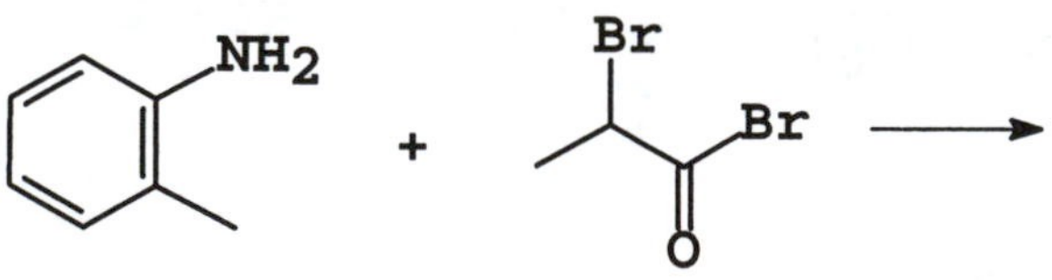

QUESTIONS 71-100

71)

72)

73)

74)

75)

76)

77)

78)

141

79)

80)

81)

82)

83)

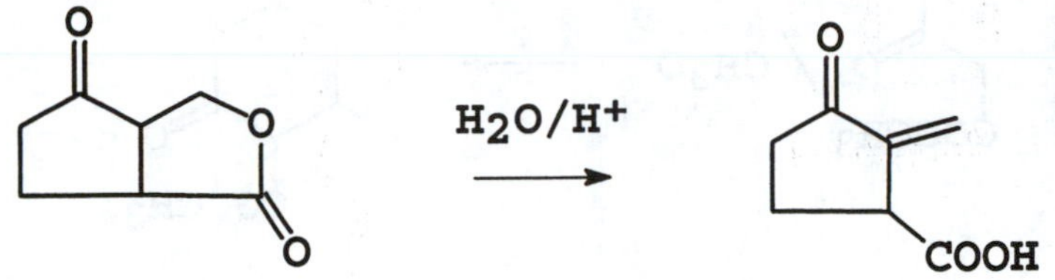

$$\xrightarrow{\text{H}_2\text{O/H}^+}$$

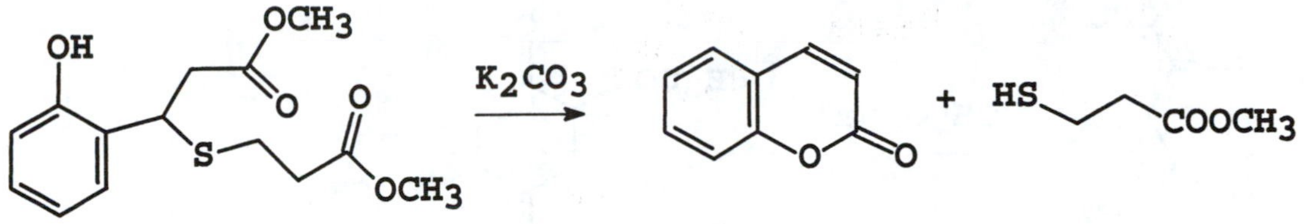

84)

$$\xrightarrow{\text{K}_2\text{CO}_3}$$

85)

$$\xrightarrow[\text{CH}_3\text{OH}]{\text{CH}_3\text{O}^-\text{Na}^+}$$

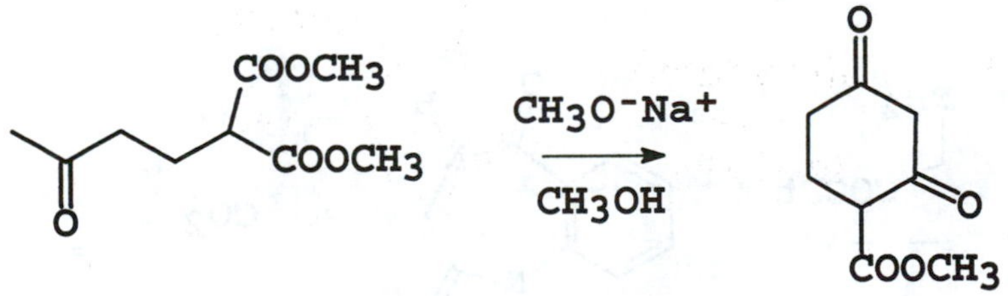

86)

$$\xrightarrow{\text{HO}^-}$$

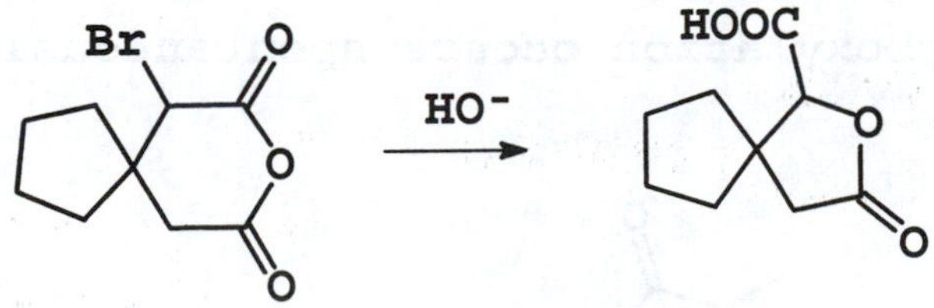

143

87)

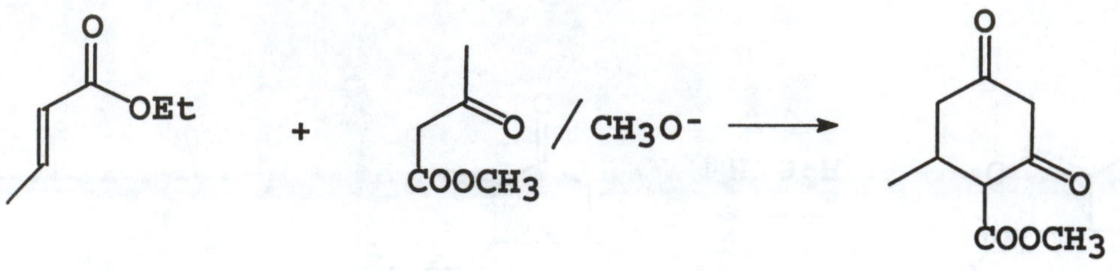

88)

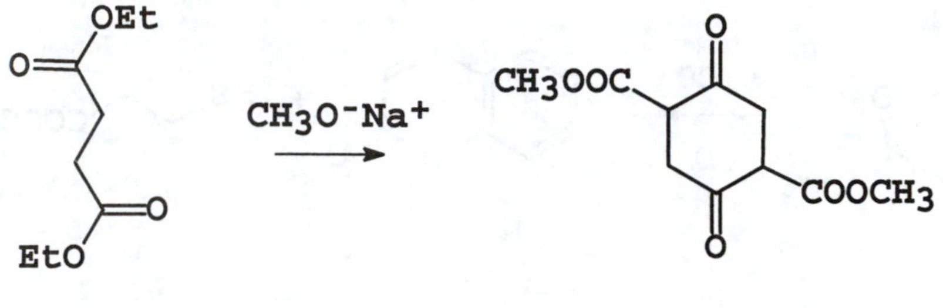

(2 mols)

89)

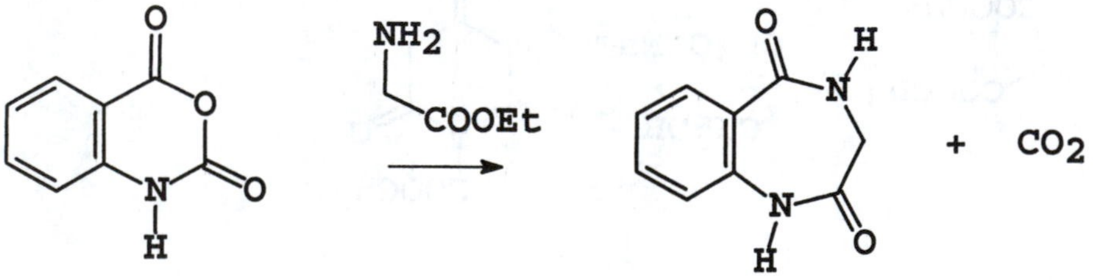

90)

The following decarboxylation occurs spontaneously.
Explain.

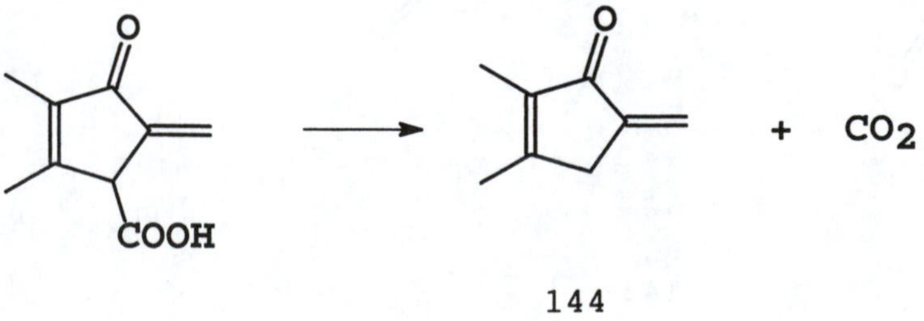

144

91)

92)

93)

94)

145

95)

96)

97)

146

98)

99)

100)

147

ANSWERS TO QUESTIONS 71-100

71)

72)

149

73)

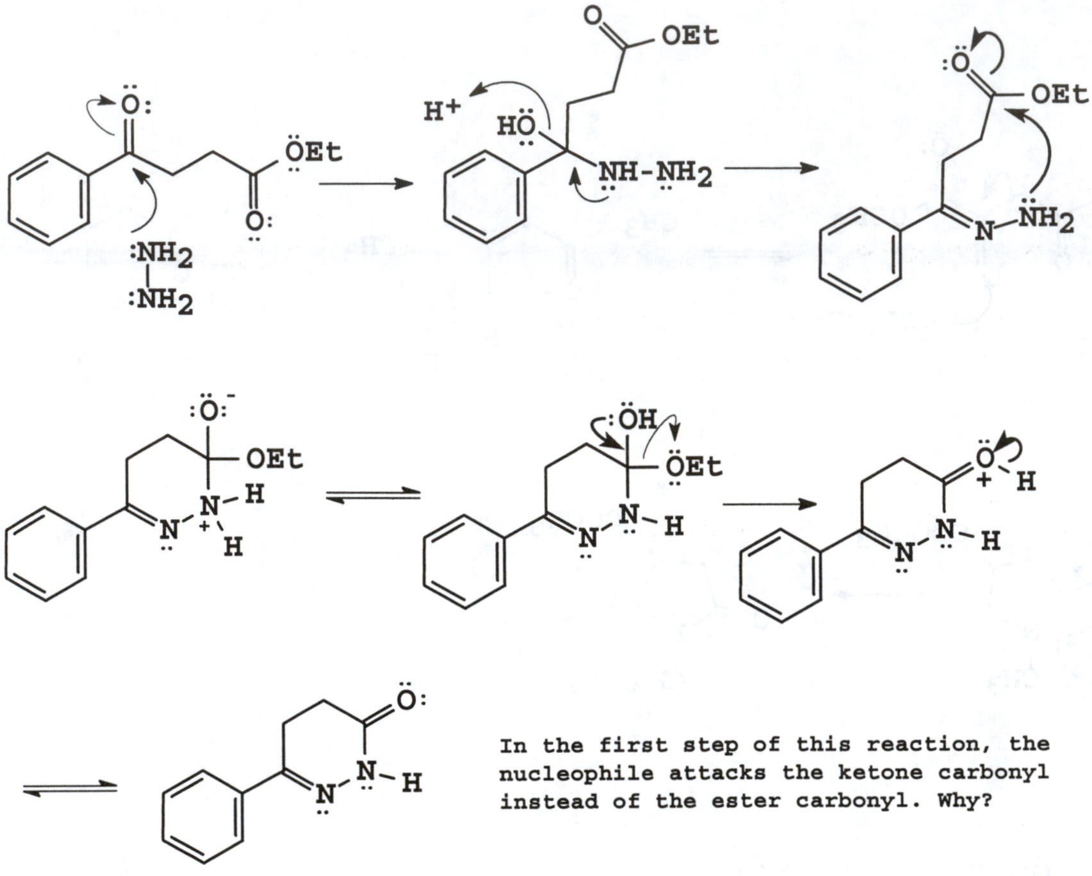

In the first step of this reaction, the nucleophile attacks the ketone carbonyl instead of the ester carbonyl. Why?

74)

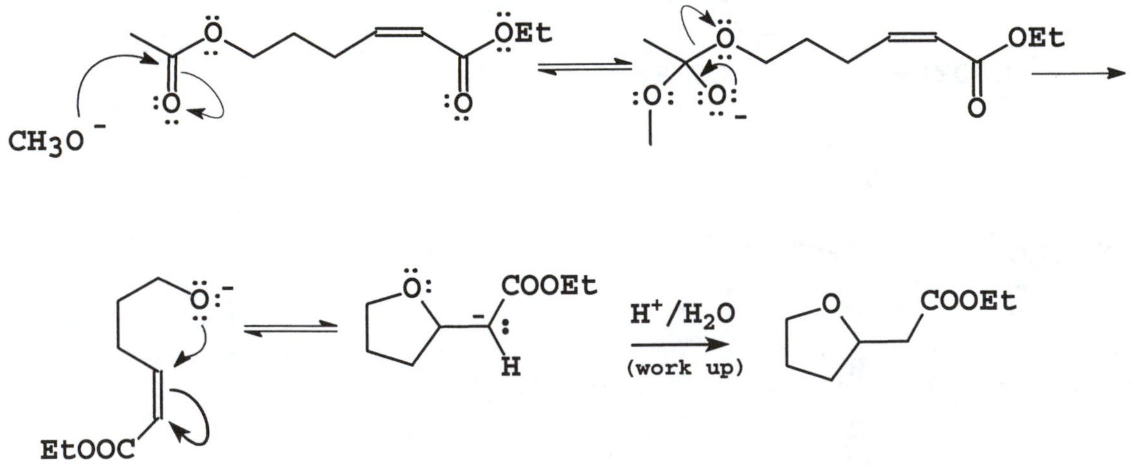

A nucleophilic acyl substitution reaction is followed by an intramolecular Michael addition reaction.

75)

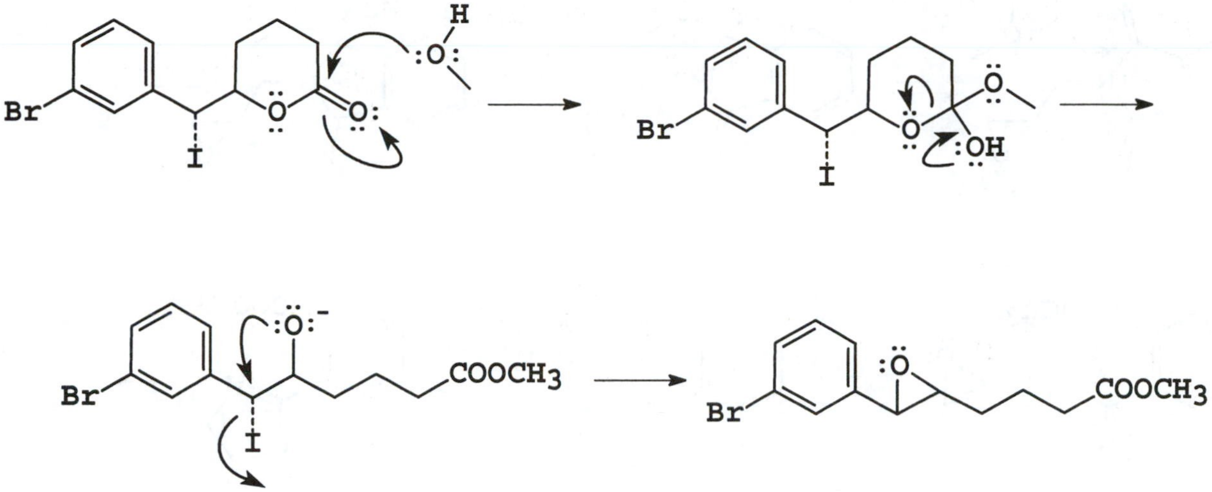

76)

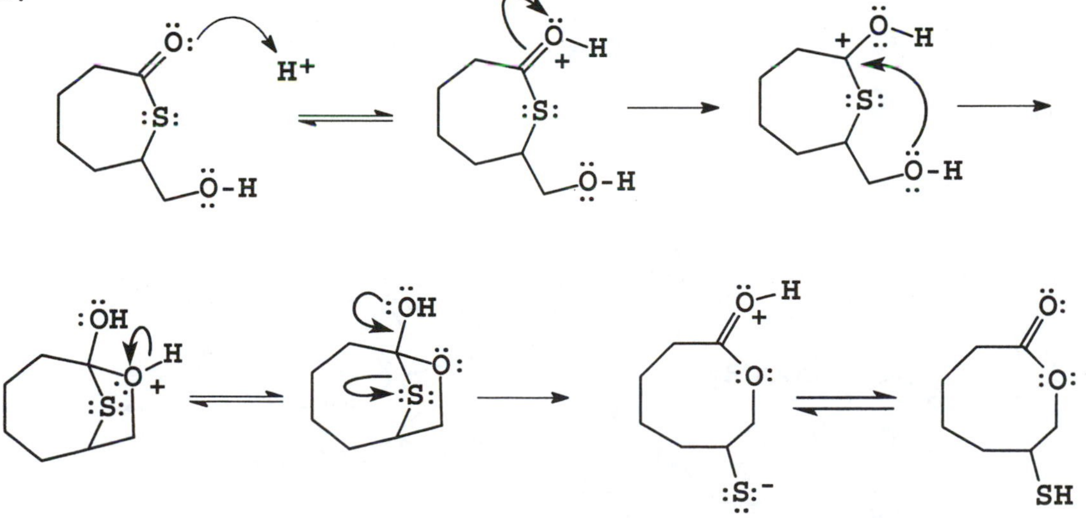

When tetrahedral intermediate A reforms the strong C=O bond, it yields B instead of the compound shown below. Why?

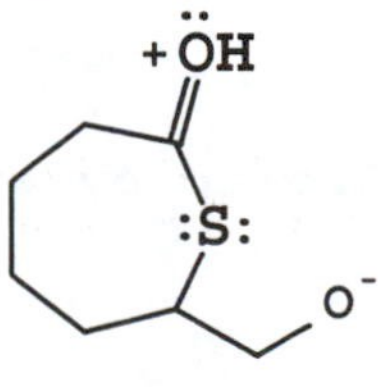

151

77)

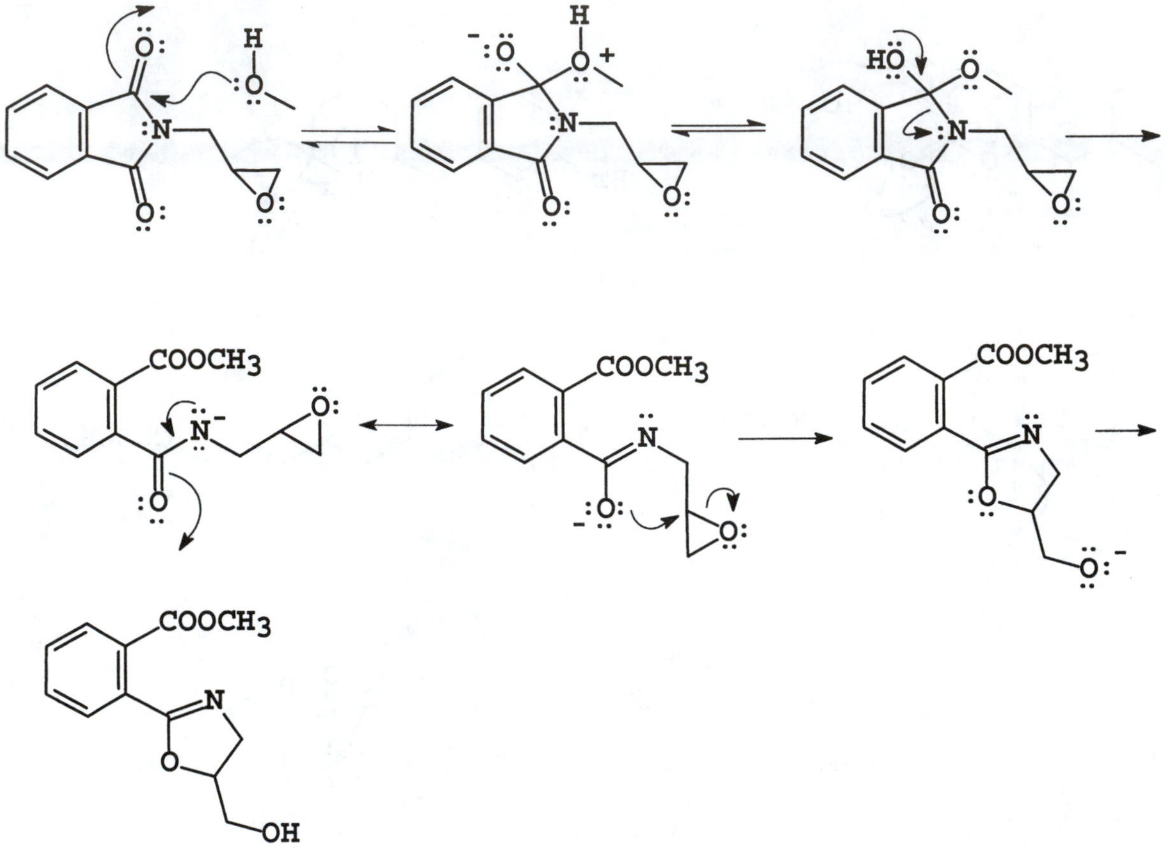

78)

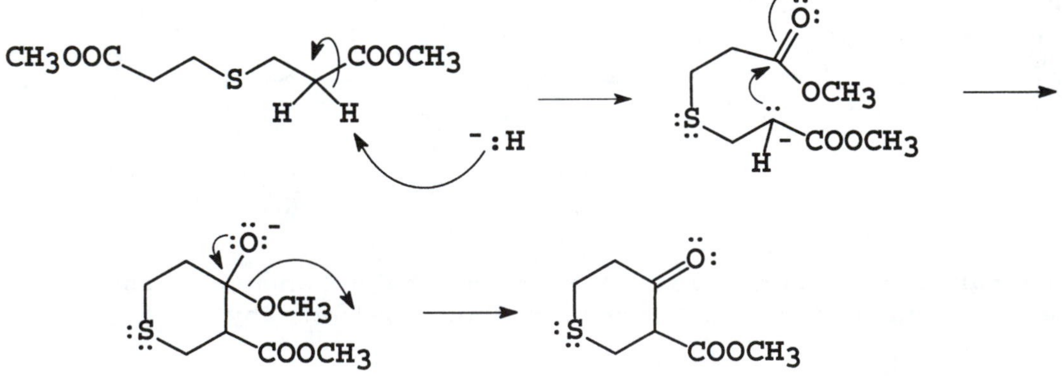

Formation of the anion is followed by an intramolecular
acyl substitution reaction.

79)

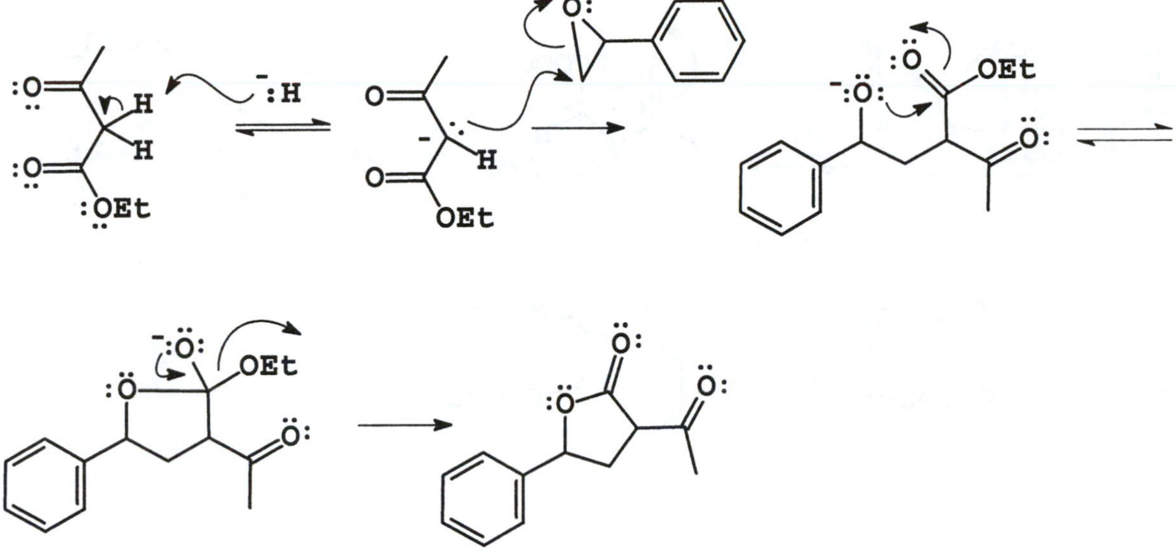

80)

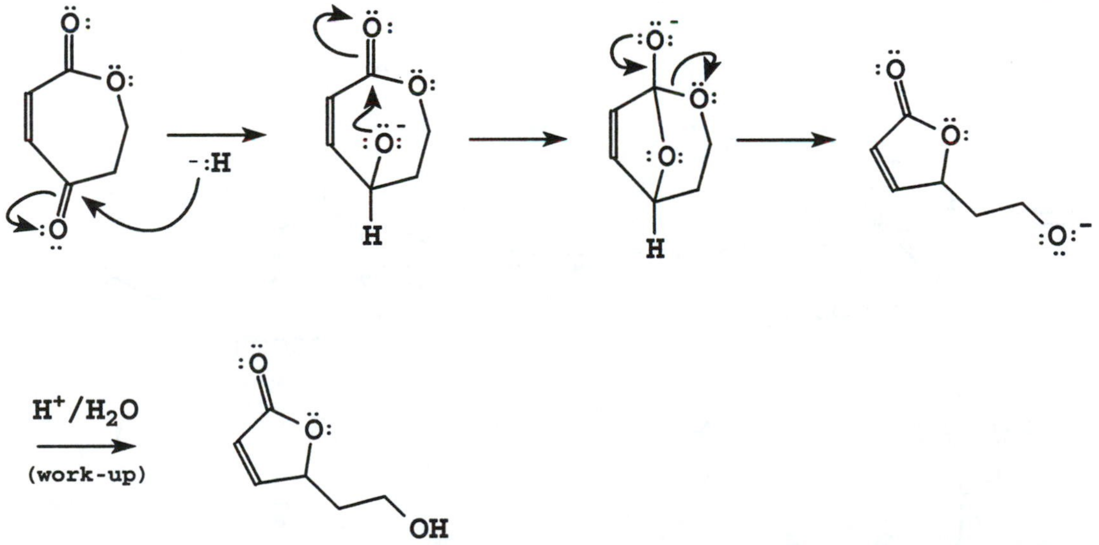

Reducing agents, such as NaBH$_4$, LiAlH$_4$, etc. function by
delivering hydride ($^-$:H) ions.

81)

82)

154

83)

84)

155

85)

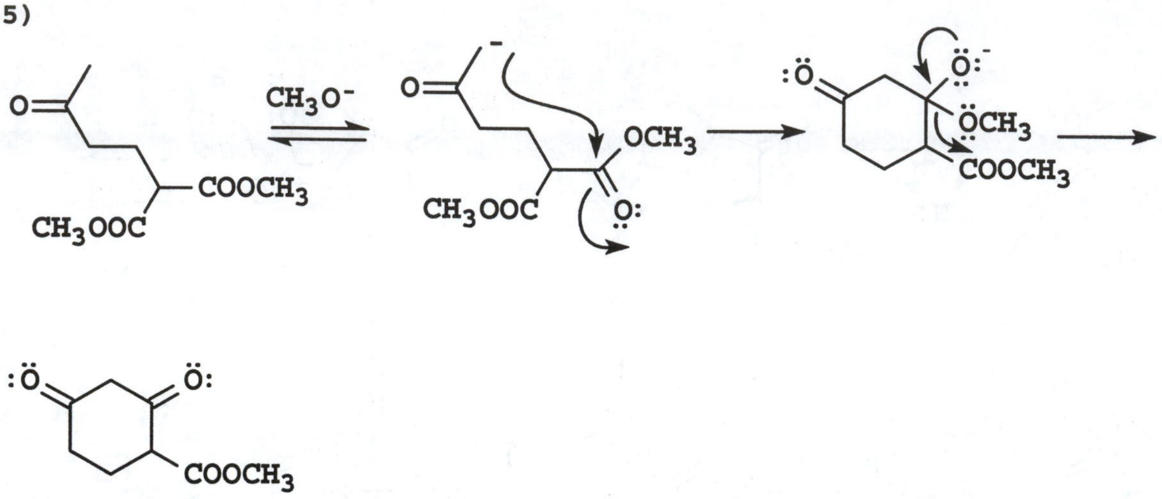

86)

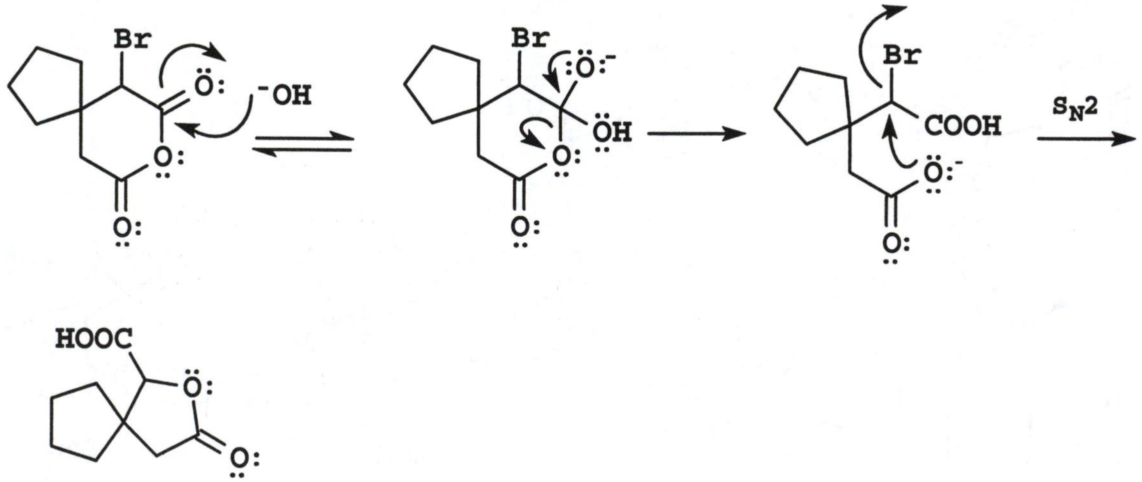

Note that in the first step of this reaction the Lewis base (nucleophile) attacks the most electron-deficient (electrophilic) carbonyl carbon. The electron-withdrawing inductive effect of the electronegative atom (Br) renders the top carbonyl group more electrophilic.

87)

88)

157

89)

NH2
COOEt
H
N+
-O
COOEt
NH
OEt
NH-
OEt

90)

H+
H
resonance-stabilized
anion

91)

HO-
HO

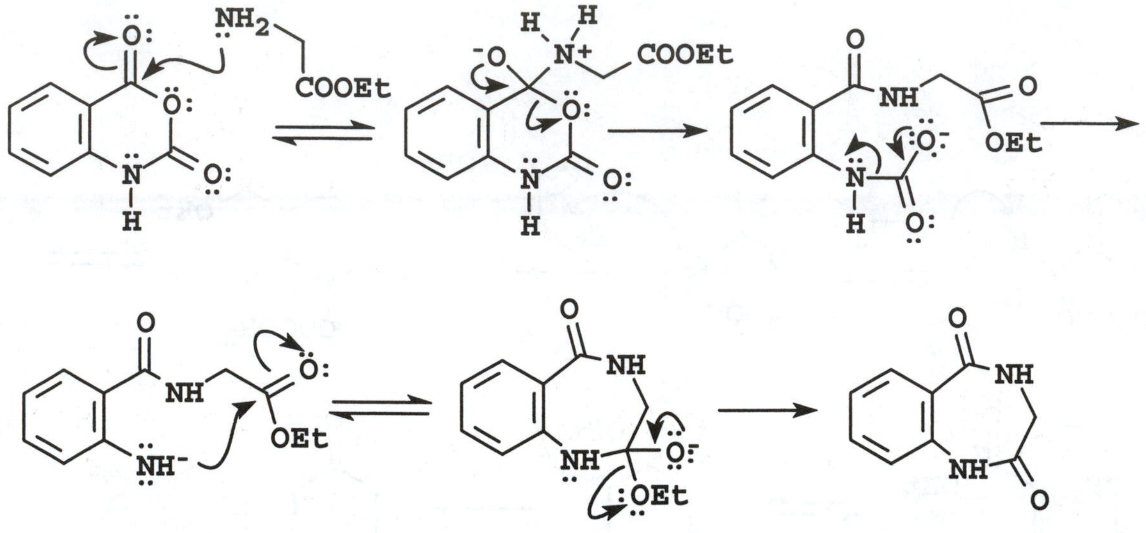

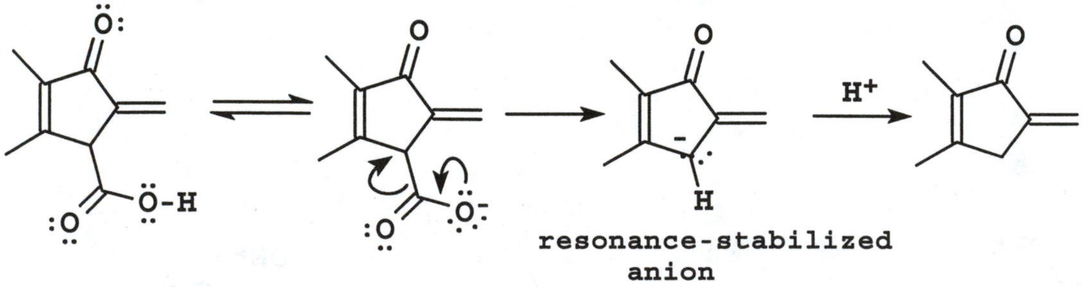

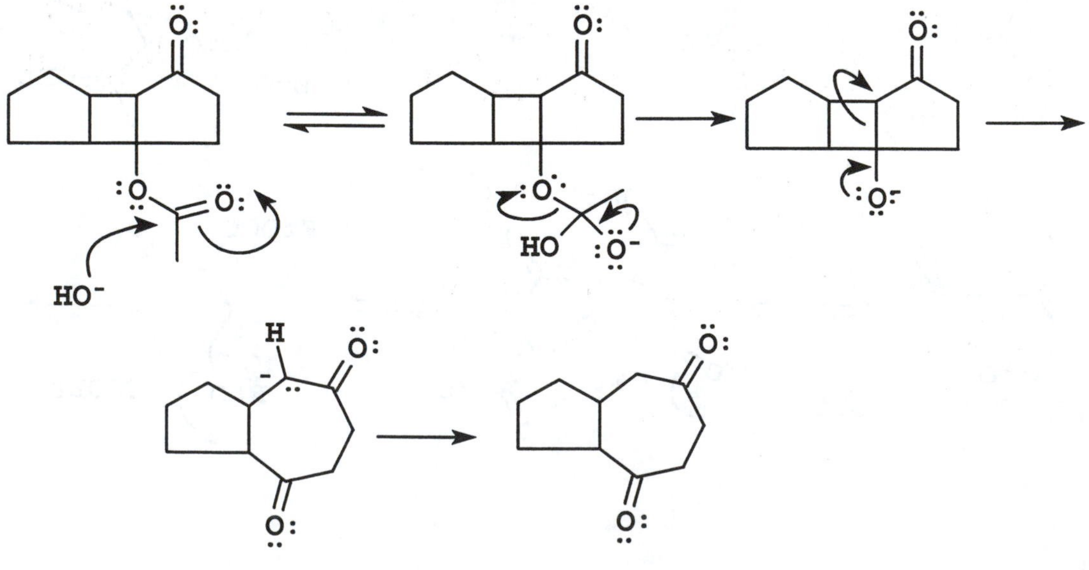

92)

93)

159

94)

95)

160

96)

97)

98)

162

99)

100)

163

PART B

Questions 101-200

101)

102)

103)

104)

105)

106)

107)

108)

166

109)

110)

111)

112)

167

113)

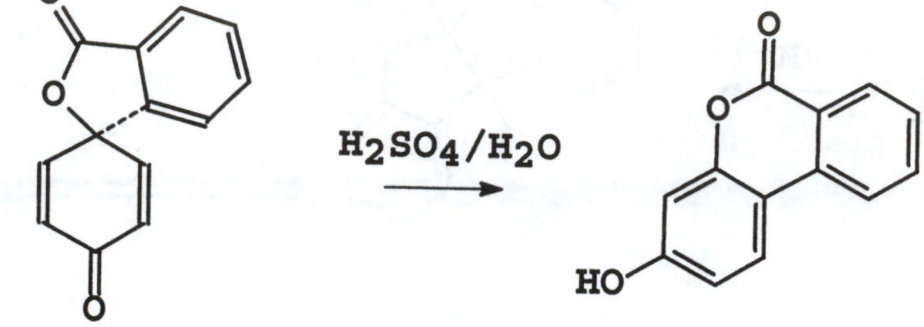

$$H_2SO_4/H_2O$$

114)

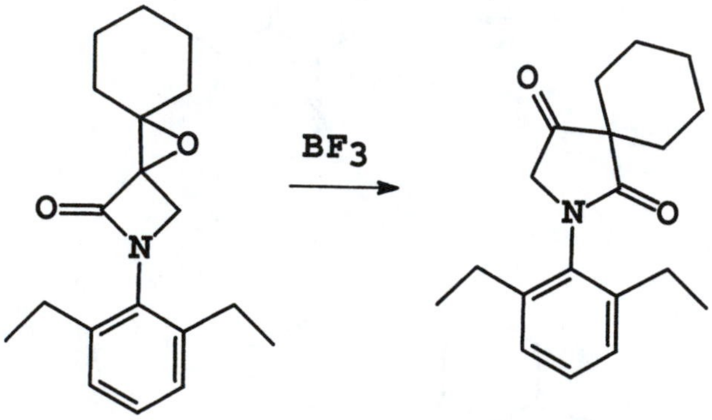

$$BF_3$$

115)

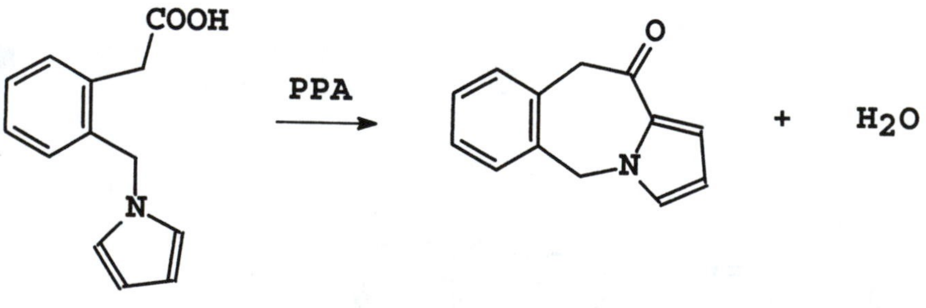

PPA

$$+ \quad H_2O$$

116)

$$+ \quad HCl$$

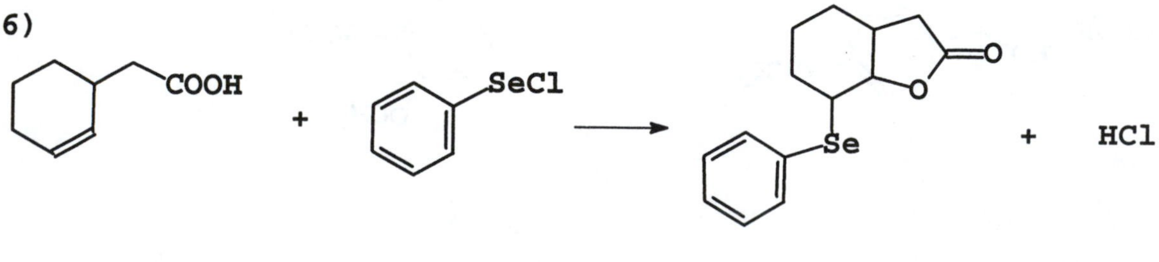

117)

118)

119)

120)

121)

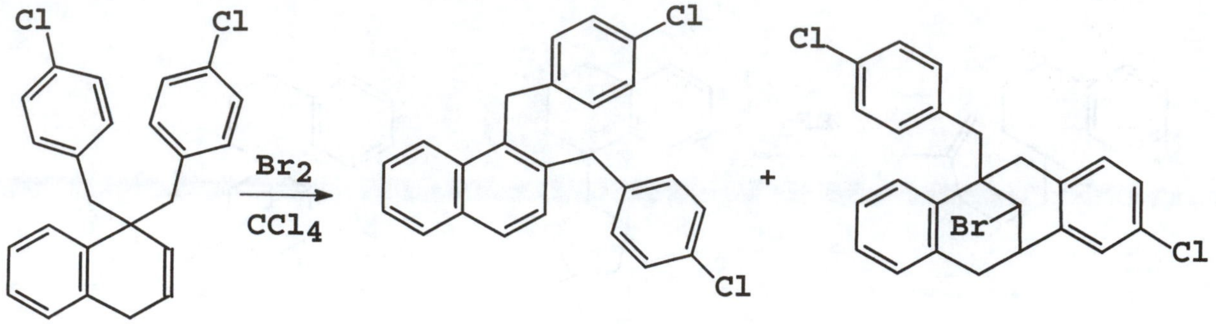

Br_2 / CCl_4

+

122)

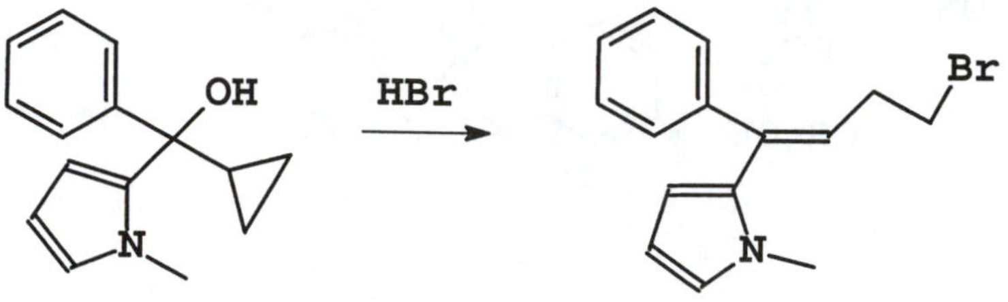

HBr

123)

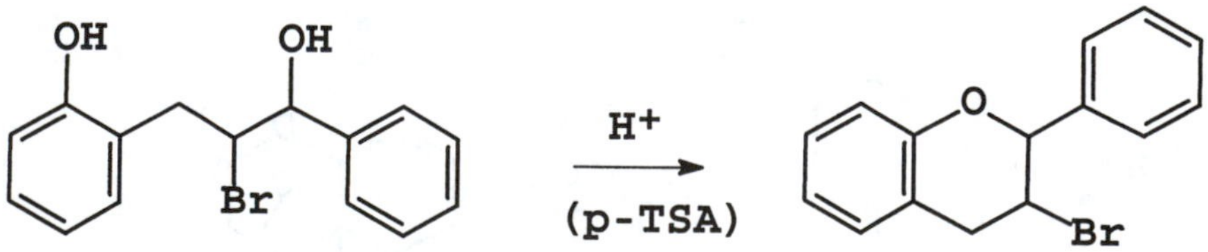

H^+ (p-TSA)

124)

CH_3O

+

Br

OH

(2 mols)

H_2SO_4

CH_3O

CH_3O

Br

Br

170

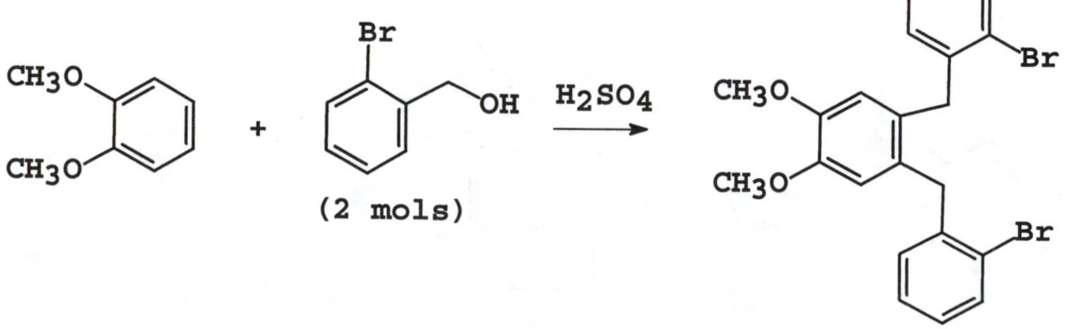

125)

126)

127)

128)

129)

130)

131)

132)

172

133)

134)

135)

136)

173

137)

138)

139)

140)

174

141)

142)

143)

144)

175

145)

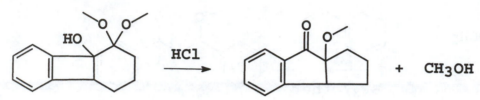

146)

147)

148)

176

149)

150)

151)

152)

153)

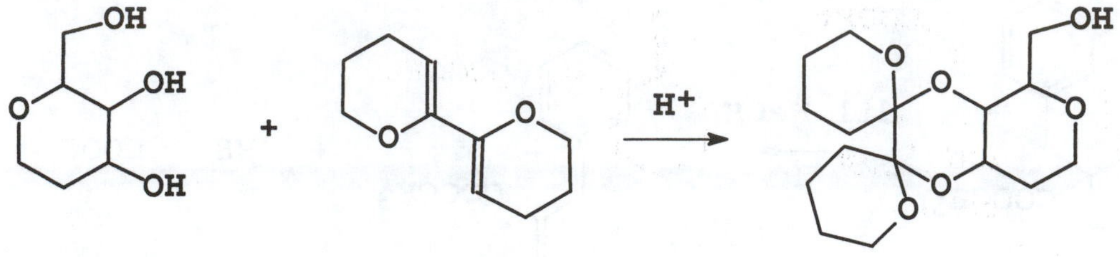

154)

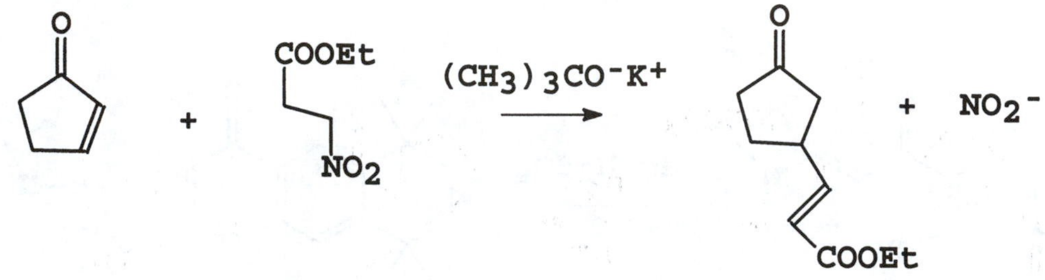

155)

The compound shown below is stable when isolated as the hydrochloride salt. However, chromatography on silica gel yields the products shown below. Write a mechanism that explains the formation of these products (<u>hint</u>: silica gel tends to absorb moisture).

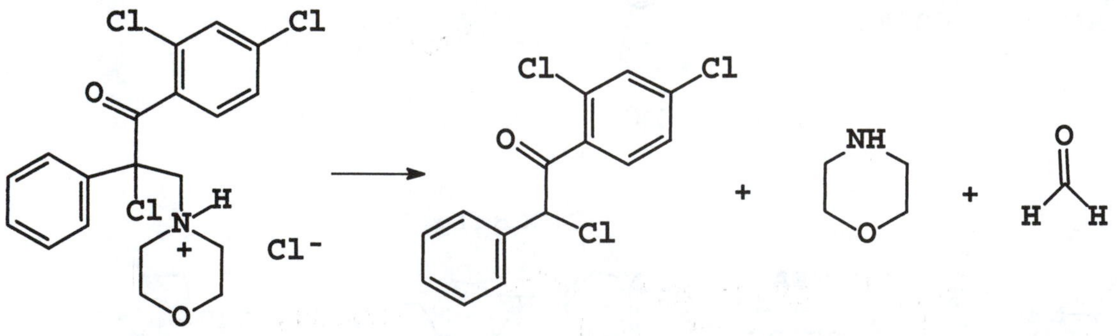

156)

157)

158)

159)

160)

161)

162)

163)

164)

165)

166)

167)

168)

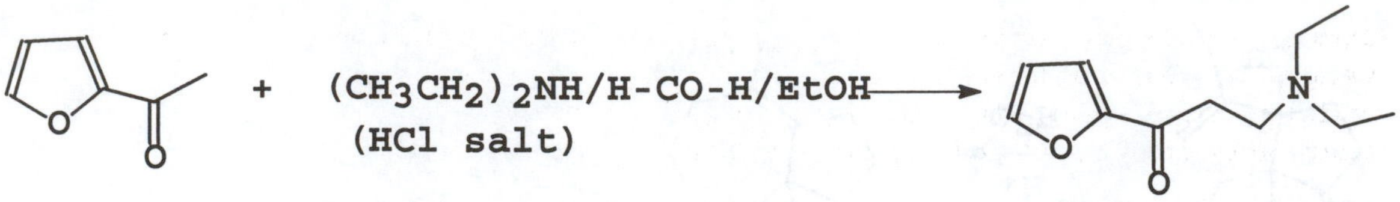

169)

Aldehydes can be converted into -chloromethyl ketones
by reacting an aldehyde with the reagent shown
below, and refluxing the product formed in toluene.
Write a mechanism for this transformation.

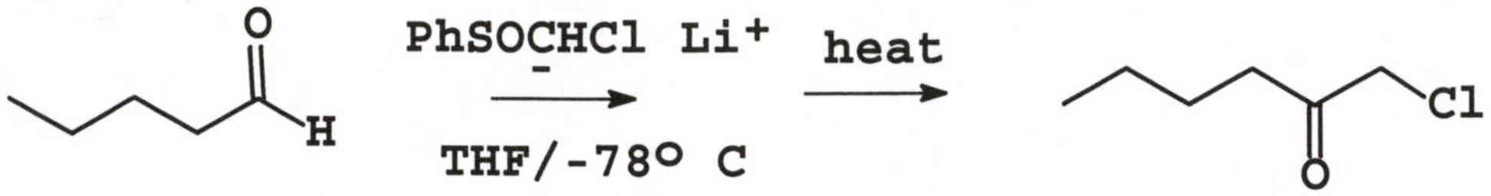

170)

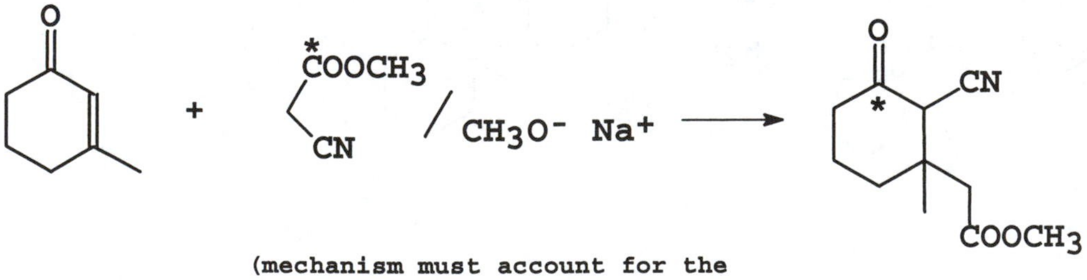

(mechanism must account for the
location of the labeled (*) carbon)

171)

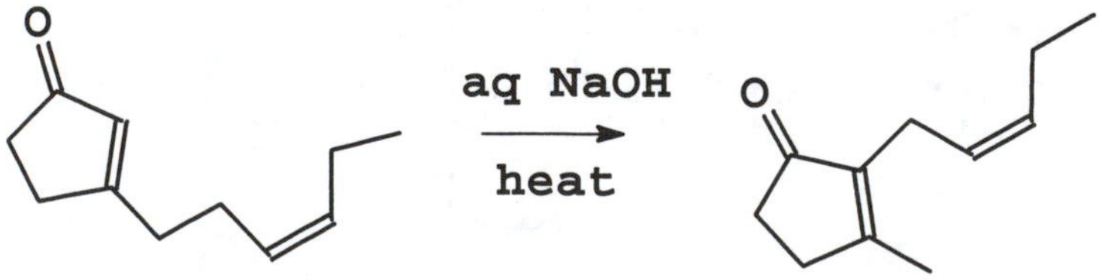

172)

Simple cyclopropanes do not react with nucleophilic reagents. In contrast, "activated" cyclopropanes undergo nucleophilic cleavage. Write a mechanism that accounts for this observation.

173)

174)

175)

176)

177)

178)

184

179)

Primary and secondary alcohols can be readily oxidized by dimethyl sulfoxide/phosphorus pentoxide/triethylamine under mild conditions (an example is shown below). Use your knowledge of Lewis structures and Lewis acid/base reactions to write a reasonable mechanism for this reaction.

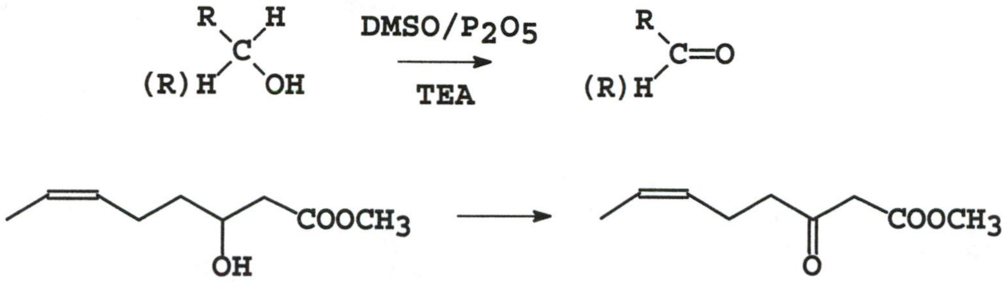

180)

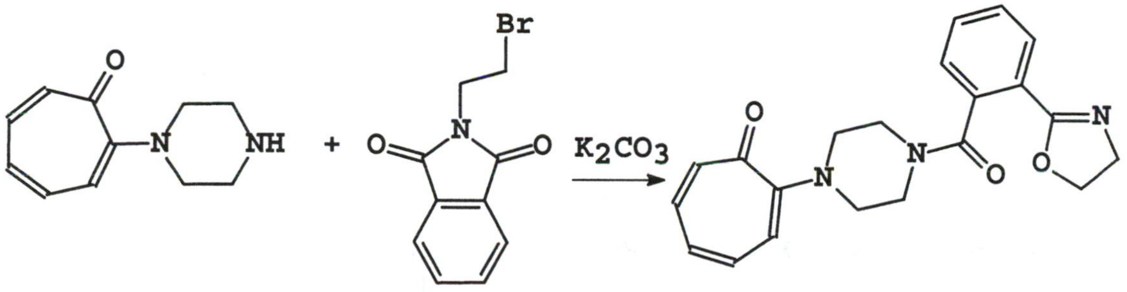

181)

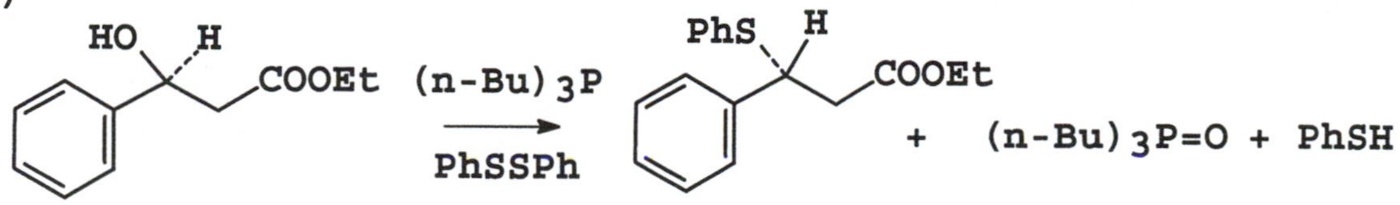

182)

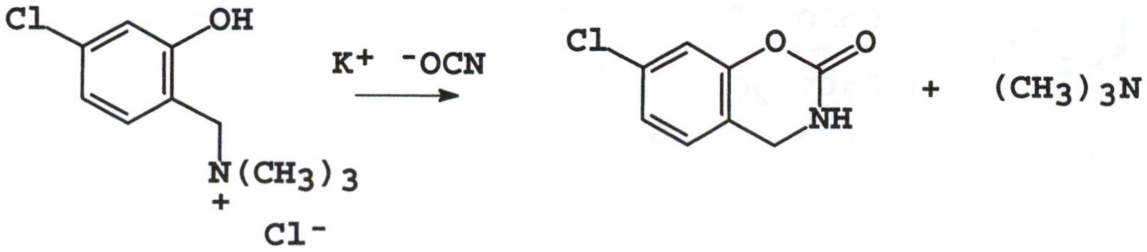

185

183)

184)

185)

186)

186

187)

Alcohols can be readily converted to alkyl chlorides or bromides under mild conditions using PPh_3/CCl_4 or PPh_3/CBr_4, respectively (a specific example is shown below). Write a mechanism for this transformation.

$$R-CH_2-OH \xrightarrow[CX_4]{PPh_3} R-CH_2X + O=PPh_3 \quad (X = Cl, Br)$$

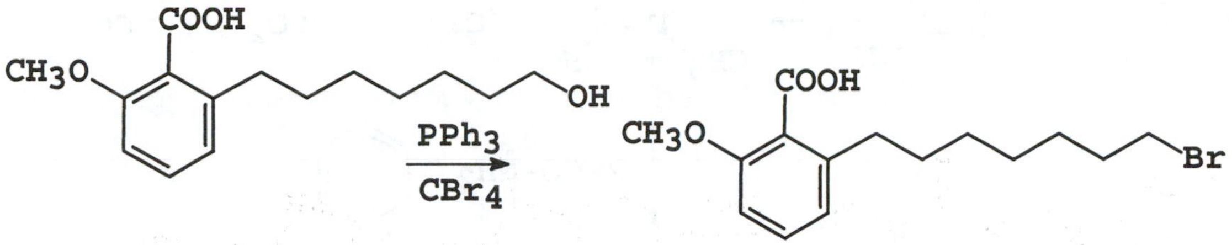

188)

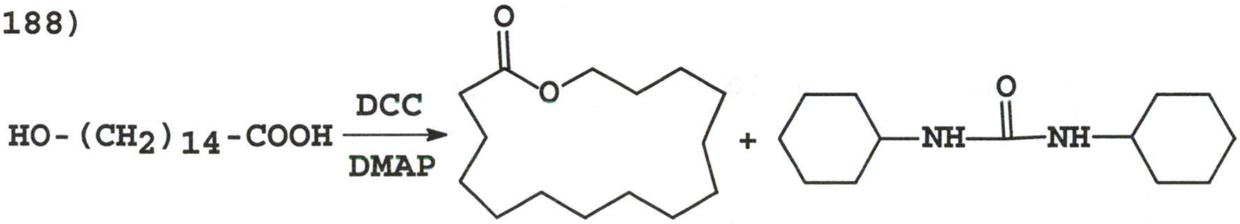

$$HO-(CH_2)_{14}-COOH \xrightarrow[DMAP]{DCC}$$

189)

$(NaCl + NaCN + 2H_2)$

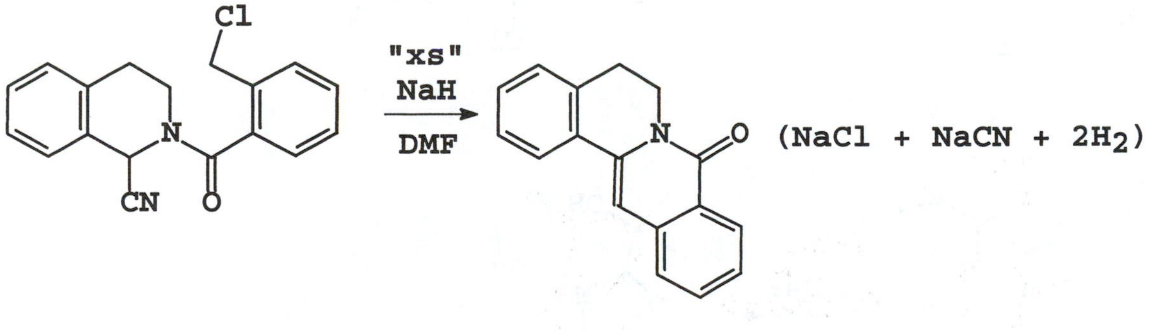

190)

Primary amides can be readily dehydrated under mild conditions using chloromethylene iminium salts (called Vilsmeier reagents) to form the corresponding nitriles. When dimethyl formamide (DMF) is treated with oxalyl chloride it forms a Vilsmeier reagent, which is then reacted with a primary amide in the presence of pyridine to yield a nitrile, as shown below. Write a <u>mechanism</u> for the formation of the Vilsmeier reagent, and a second <u>mechanism</u> for the formation of the nitrile.

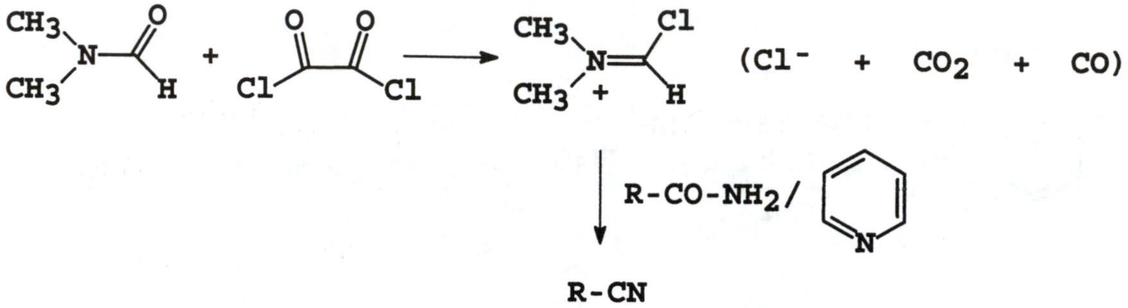

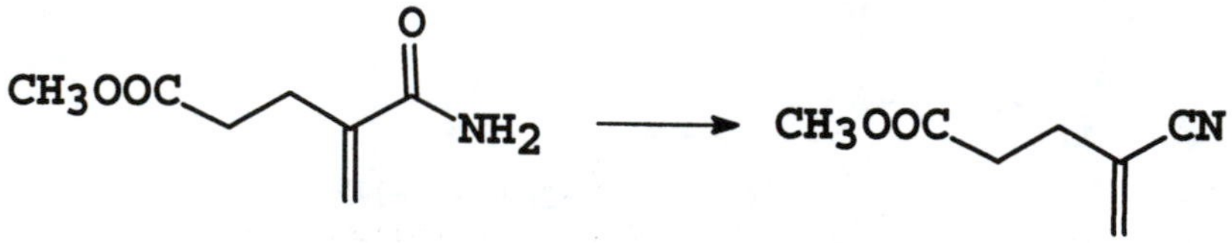

191)

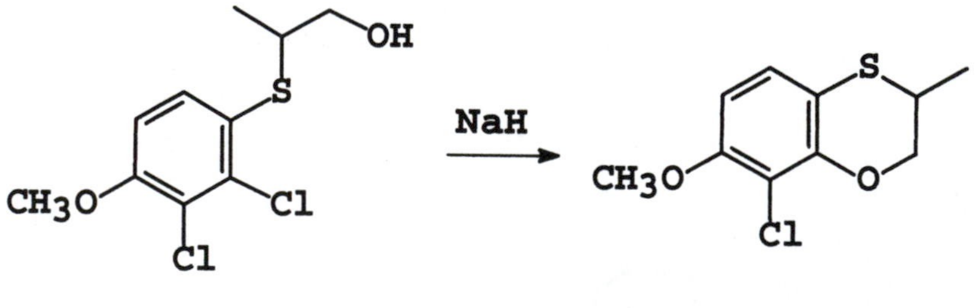

192)

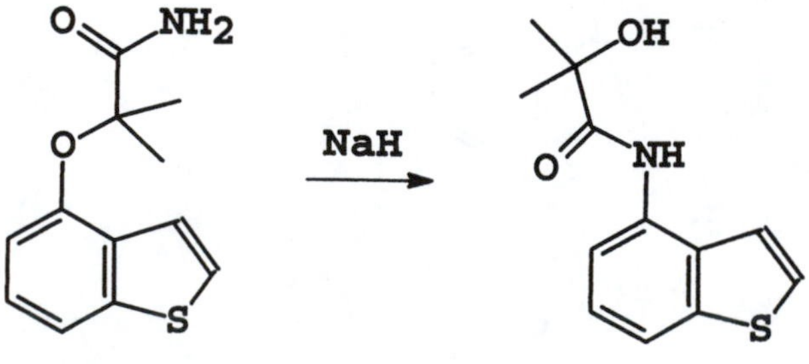

193)

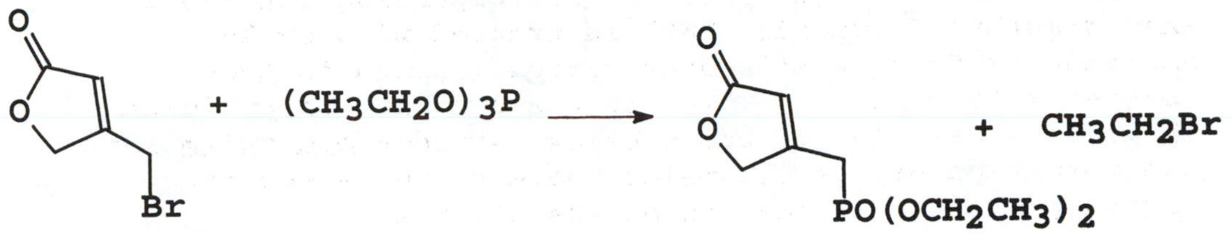
O
O
Br
+
(CH3CH2O)3P
O
O
PO(OCH2CH3)2
+
CH3CH2Br

194)

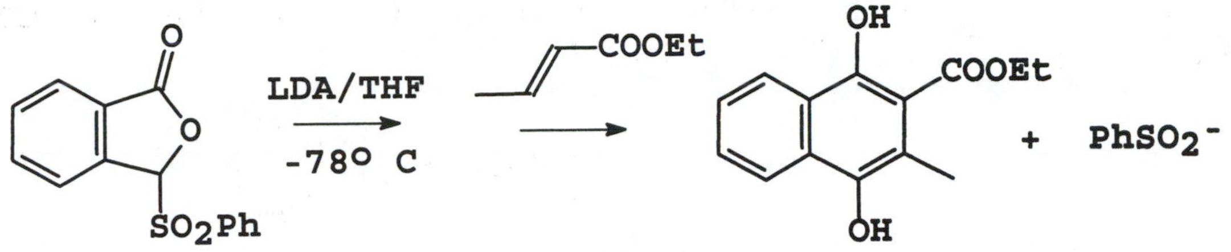
O
O
SO2Ph
LDA/THF
-78° C
COOEt
OH
COOEt
OH
+
PhSO2-

195)

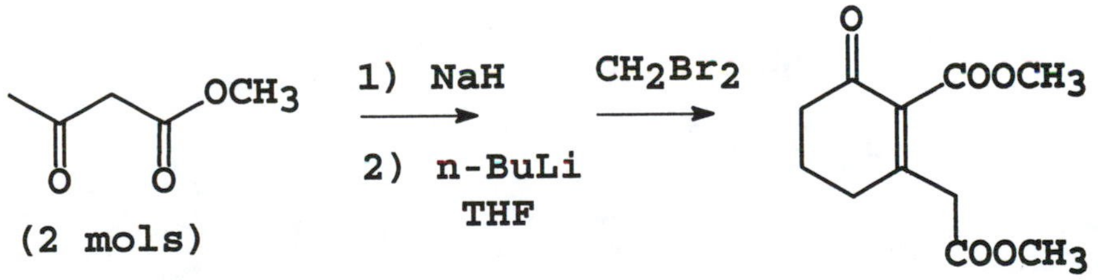
OCH3
O
O
(2 mols)
1) NaH
2) n-BuLi
THF
CH2Br2
O
COOCH3
COOCH3

196)

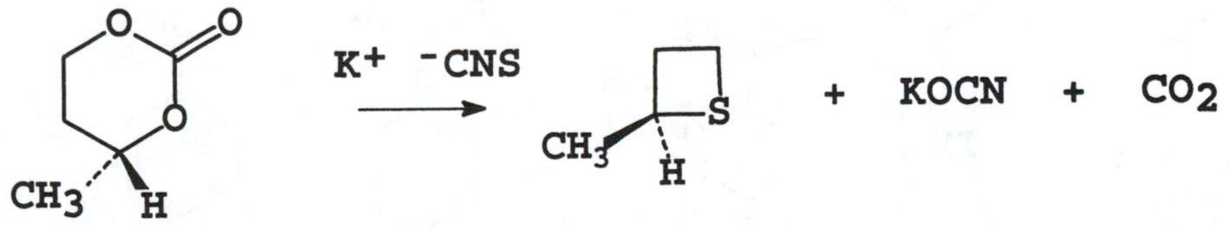
O
O
O
CH3
H
K+ -CNS
S
CH3
H
+
KOCN
+
CO2

197)

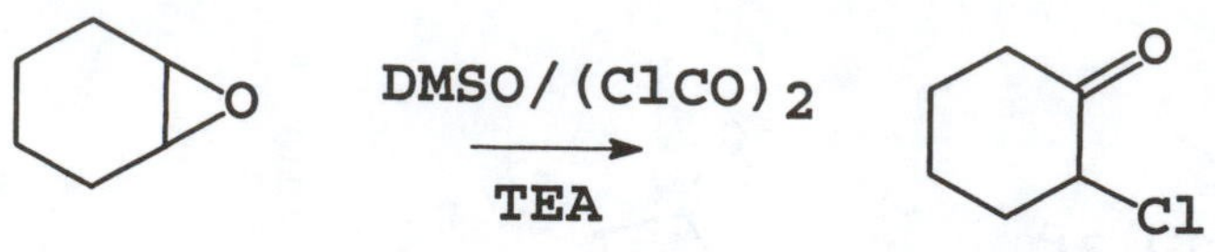

DMSO/(ClCO)$_2$

TEA

198)

Ph$_3$P/DEAD

PhSH

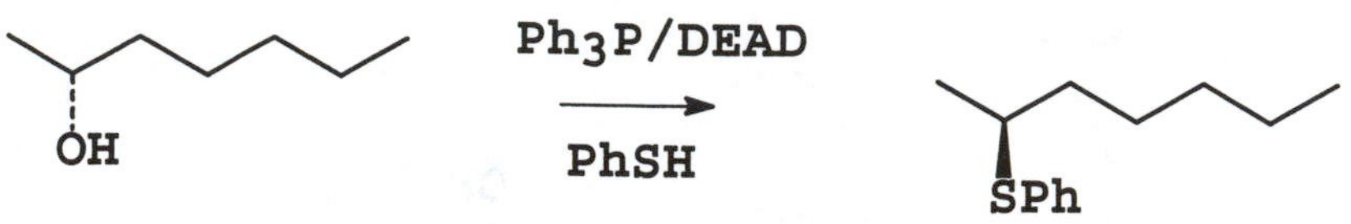

199)

t-Butyl ethers and esters can be readily made from the corresponding alcohols and carboxylic acids by mixing the reagent shown below with t-butanol in the presence of a catalytic amount of BF$_3$ etherate. Write a mechanism for these tranformations.

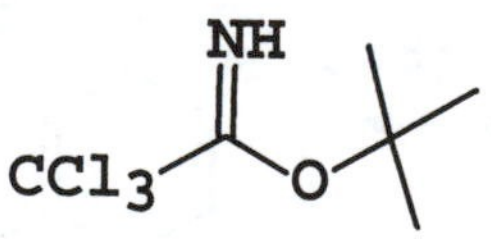

200)

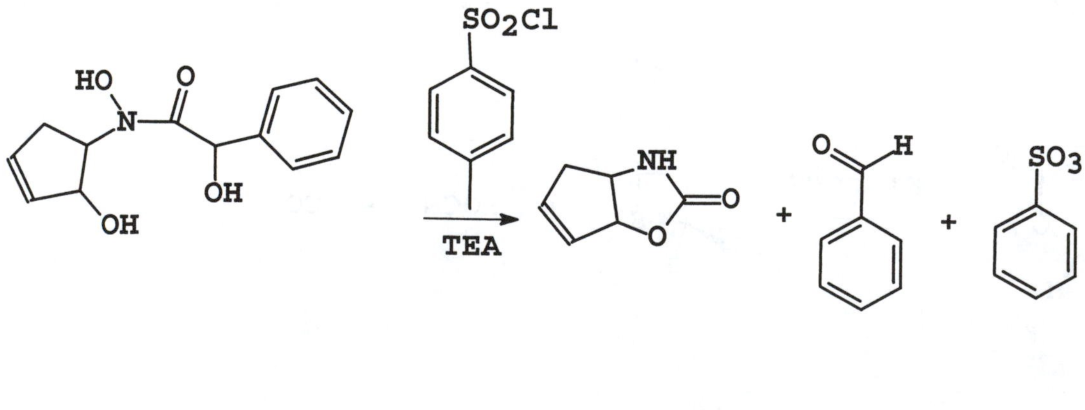

SO$_2$Cl

TEA

SO$_3^-$

190

ANSWERS TO QUESTIONS 101-200

101)

102)

103)

193

104)

105)

194

106)

107)

195

108)

109)

196

110)

111)

197

112)

113)

198

114)

115)

116)

117)

200

118)

119)

201

120)

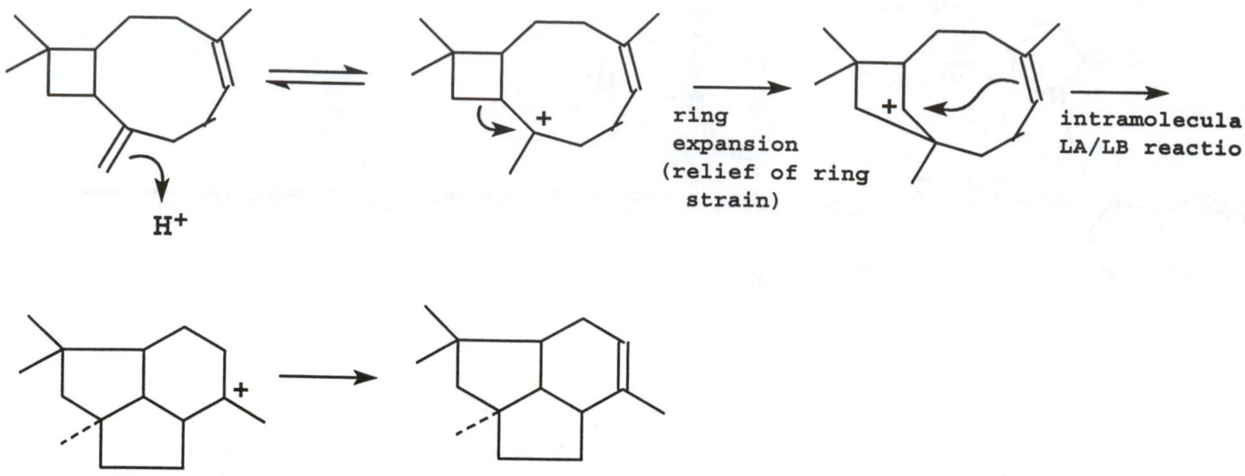

H+
ring
expansion
(relief of ring
strain)
intramolecula
LA/LB reactio

121)

Ar Ar
Br—Br
Br+
(I)
Br
Ar
Ar
Br
H H
Br-
Ar Ar
Br
(I)
Ar
Br
Ar
Br
H
Cl

Ar
Br
Cl

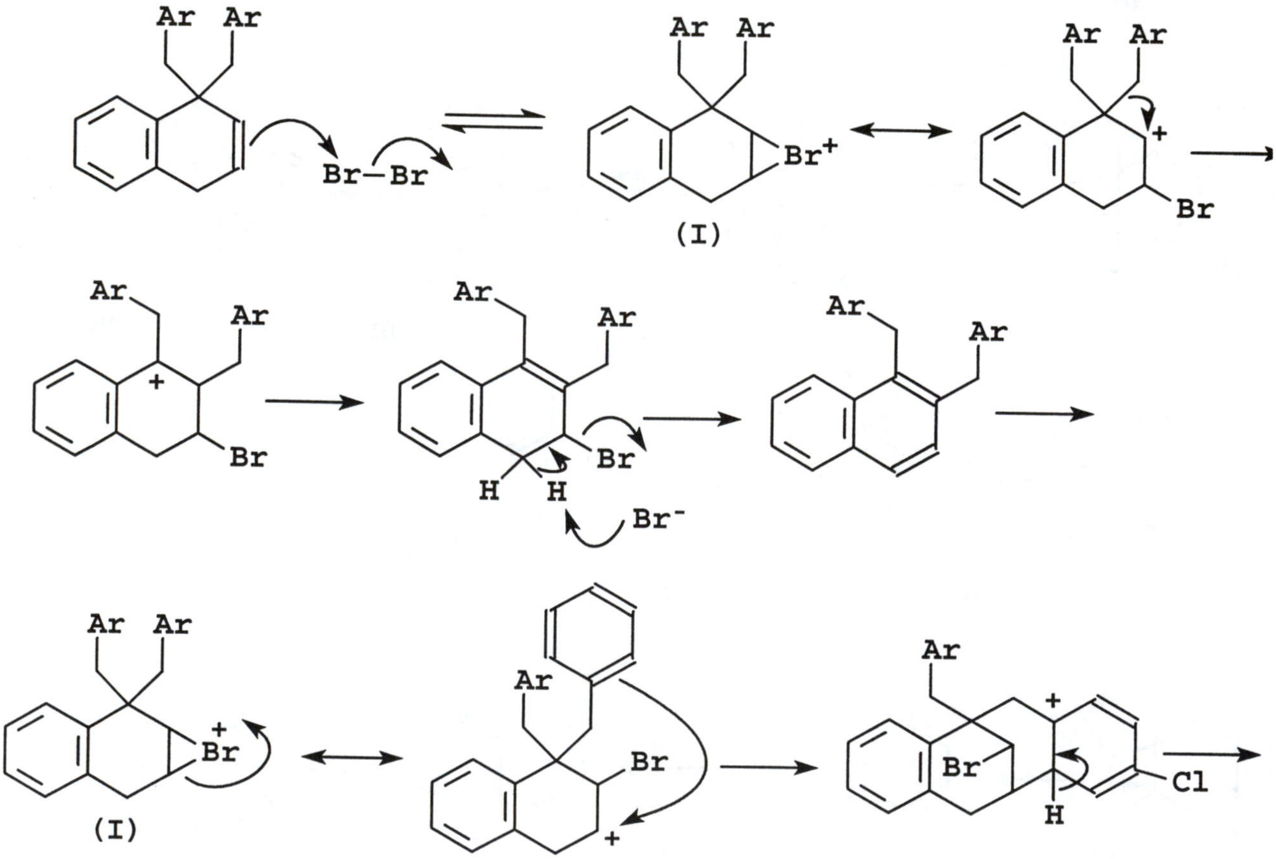

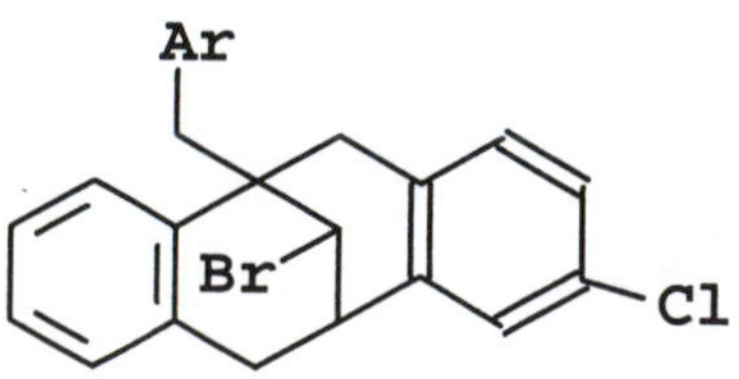

122)

123)

203

124)

125)

126)

127)

205

128)

129)

206

130)

131)

207

132)

133)

208

134)

135)

209

136)

137)

210

138)

139)

211

140)

141)

142)

143)

213

144)

145)

214

146)

147)

215

148)

149)

216

150)

151)

217

152)

153)

218

154)

155)

156)

157)

158)

159)

221

160)

161)

162)

163)

223

164)

165)

224

166)

167)

225

168)

169)

170)

171)

227

172)

resonance-stabilized
anion

173)

174)

175)

229

176)

(+ DMAP)

177)

178)

179)

231

180)

181)

232

182)

183)

233

184)

185)

234

186)

187)

188)

189)

236

190)

191)

237

192)

193)

194)

195)

196)

197)

198)

199)

241

200)

PART C

Questions 201-210

201) When the antibiotic emycin F is treated with acid, it rearranges to emycin E. Write a __mechanism__ for this transformation.

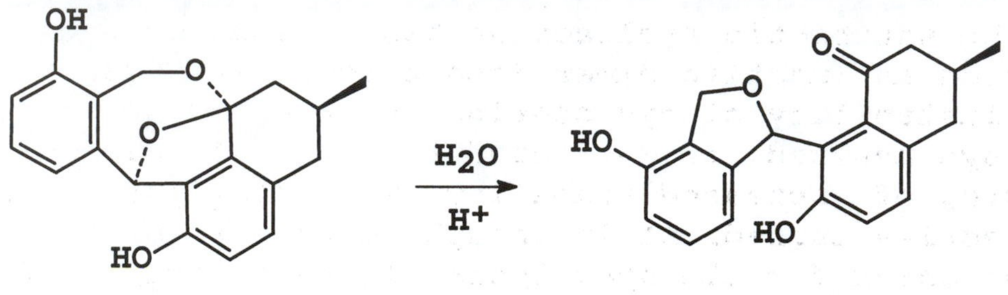

emycin F   emycin E

202) Mechlorethamine hydrochloride (Mustargen) is a drug used in the treatment of Hodgkins disease and lymphosarcoma. Mustargen is administered intravenously, and its action lasts for only a few minutes. Its short duration of action is due to its rapid non-enzymatic hydrolysis. Write a mechanism that accounts for its rapid hydrolysis (recall that ordinary alkyl halides do not react with weak nucleophiles, such as water).

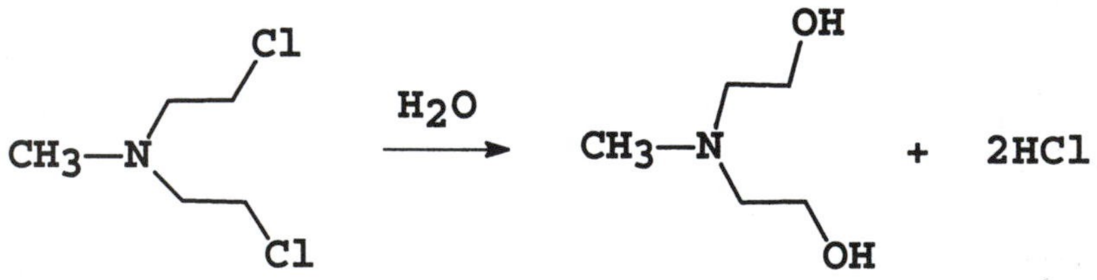

(HCl salt)

203)   A prodrug is a biologically inactive form of a drug
that in vivo yields the active form of the drug. The
liberation of the drug from the prodrug can be a
nonenzymatic or enzyme-catalyzed process. Prodrugs
are typically used because of their greater chemical
stability, better transport and/or lower toxicity.

The antibiotic cycloserine has a tendency to dimerize,
forming an inactive dimer (see below). In order to lower
the instability of cycloserine the prodrug shown below
was synthesized. It was found to be an efficacious
prodrug of increased stability. The prodrug releases
the active ingredient in phosphate buffer, pH 7.0. Write
a mechanism for the hydrolysis of the prodrug to form
cycloserine and a mechanism for the formation of the
cycloserine dimer.

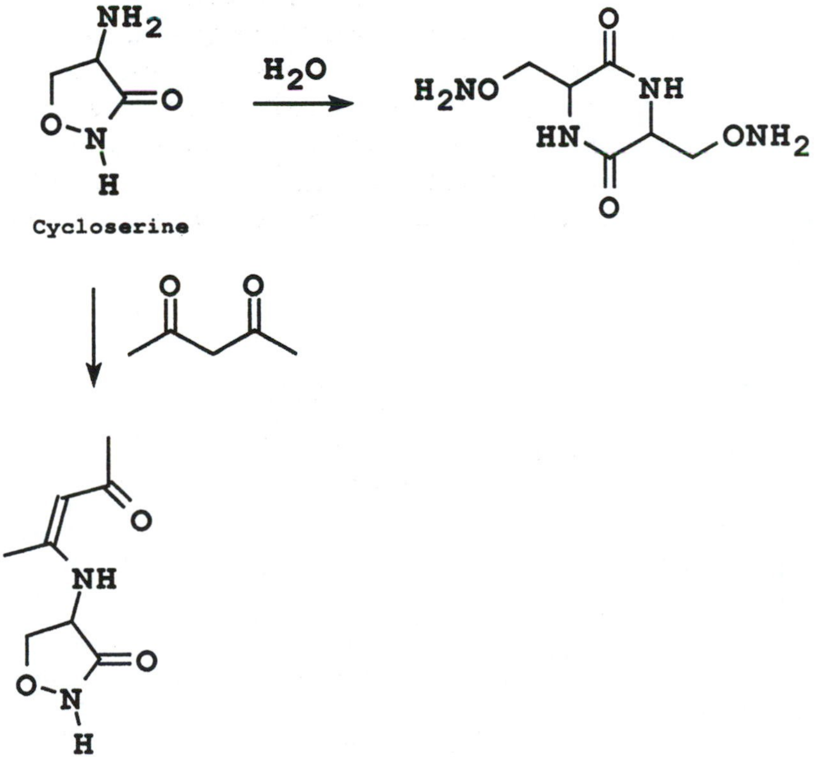

Cycloserine

204)   Mammalian cells produce a range of bioactive substances, including the prostaglandins, prostacyclins, thromboxanes and leukotrienes. These hormone-like substances act as mediators of inflammation, pain, fever, blood clotting, etc.

Prostacyclin is a potent vasodilator and inhibitor of platelet aggregation produced by vascular endothelial cells. Decreased production of prostacyclin is associated with platelet aggregation and the formation of blood clots (thrombosis). The latter is the primary cause of heart attacks and stroke. Prostacyclin is transformed into the compound shown below _in vivo_. Write a _mechanism_ for this reaction.

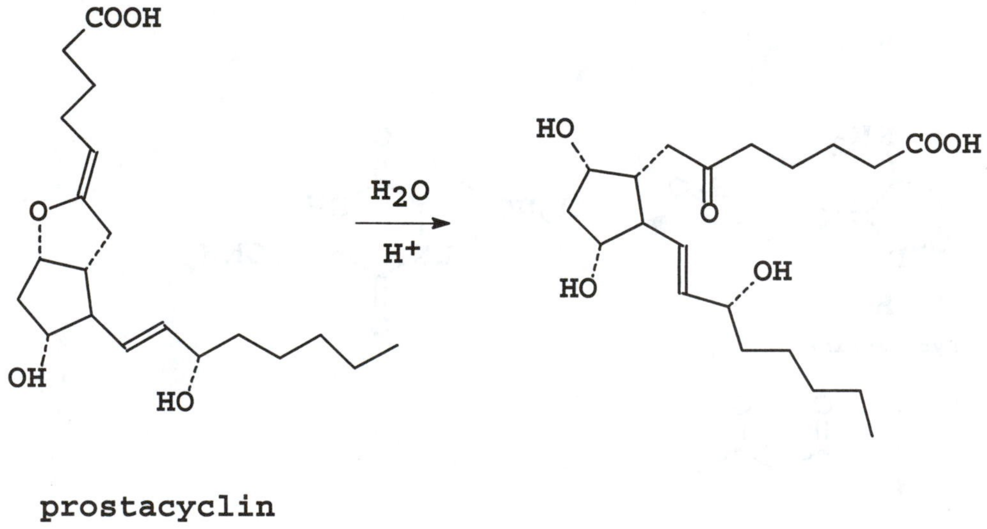

prostacyclin

205) The antitumor antibiotic leinamycin cleaves DNA, however, it does so only in the presence of added thiols. The precise mechanism of action of leinamycin is not known. It has been suggested that nucleophilic attack of a thiol on the 1,2-dithiolan-3-one 1-oxide heterocle present in leinamycin triggers DNA cleavage.

The thiol-activated DNA-cleavage chemistry of leinamycin was probed using 1,2-dithiolan-3-one 1-oxide as a model compound and n-propane thiol (see reaction below). Write a plausible mechanism for this reaction.

leinamycin

206) The prodrug shown below yields the active ingredient (chloral, a sedative) in the stomach. Write a mechanism for this transformation.

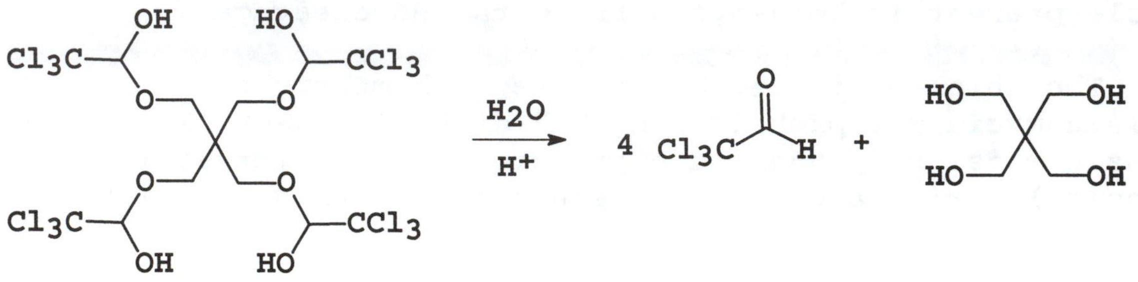

207) Solutions of the broad-spectrum antibiotic chlortetracycline hydrochloride lose their therapeutic potency with time. This is due to rapid epimerization at the C-4 position, forming an epimerized product having greatly reduced antibiotic activity. Write a mechanism for the epimerization reaction.

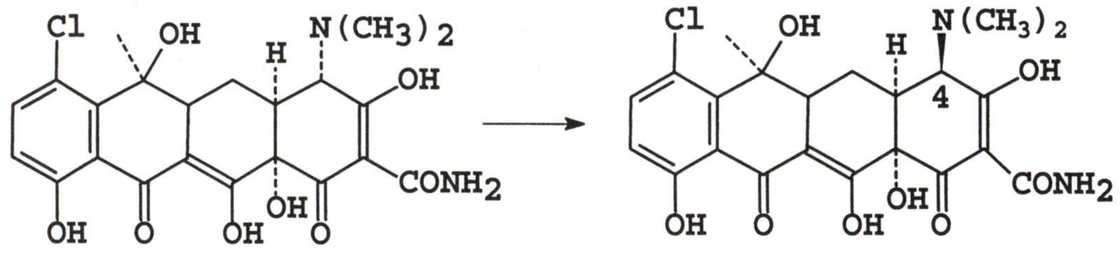

chlortetracycline
hydrochloride
(Aureomycin)

208 The intracellular enzyme CMK-KDO synthetase is the key
     enzyme in the biosynthesis of the lipopolysaccharide (LPS)
     of gram-negative bacteria. Compounds that inhibit this
     enzyme are of potential therapeutic value as antibacterial
     agents.

Compound 1 is a potent inhibitor of CMP-KDO synthetase
in vitro but is inactive as an antibacterial agent in vivo.
In an effort to circumvent this problem a series of simple
esters were synthesized and investigated for their antibacterial activity. None of the esters showed any activity.
Ultimately, compound 2 was conceived which was found to be
a highly effective antibacterial agent. The design of
prodrug 2 was based on the following biological and
chemical considerations:

(a) gram-negative bacteria contain significant quantities
of glutathione (γ-asp-cys-gly);
(b) in vitro disulfide exchange reactions are facile;
(c) the disulfide exchange was expected to be
irreversible because of the "peri" effect;
(d) when model compound 3 was reacted with n-propanethiol
in the presence of triethylamine at room temperature,
compound 4 ($C_{11}H_8S$; ^1H NMR: 4.78 (s,2H), 7.20-7.60
(m, 4H), was formed.

Write a mechanism for the in vivo conversion of 2 to 1,
and comment on the biochemical and chemical rationale
underlying the design of compound 2.

1

2

3

209) Isoilludin M behaves as a bifunctional alkylating agent, yielding the aromatic product shown below. Write a plausible mechanism for this reaction.

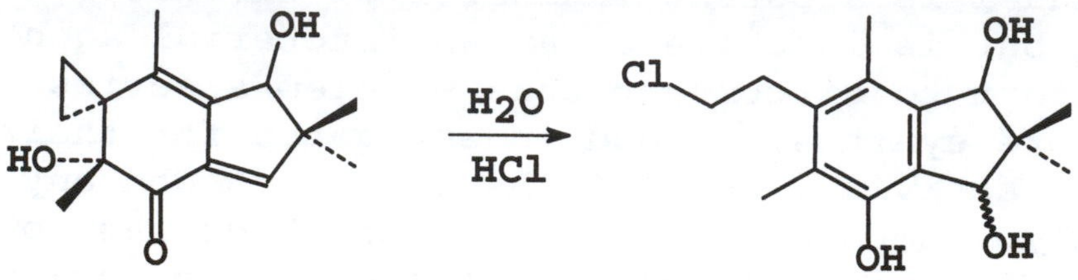

isoilludin M

210) The natural product Bripiodionen is an inhibitor of human cytomegalovirus protease. When dissolved in methanol for a prolonged period of time, bripiodionen undergoes geometric isomerization to compound X. The chemical shift of H-8 in bripiodionen is 7.45 and 7.61 in compound X. What is the structure of X? Write a mechanism for the formation of X.

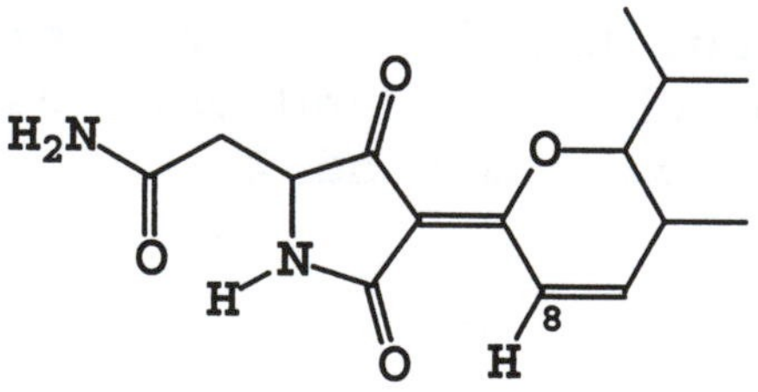

Bripiodionen

ANSWERS TO QUESTIONS 201-210

201)

202)

203)

253

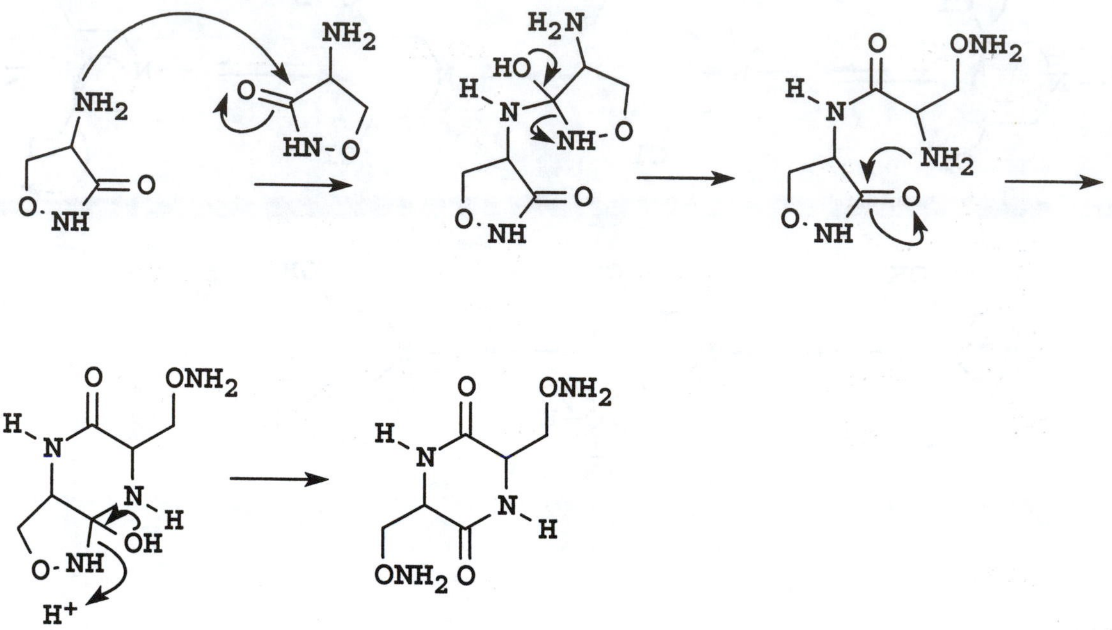

204)

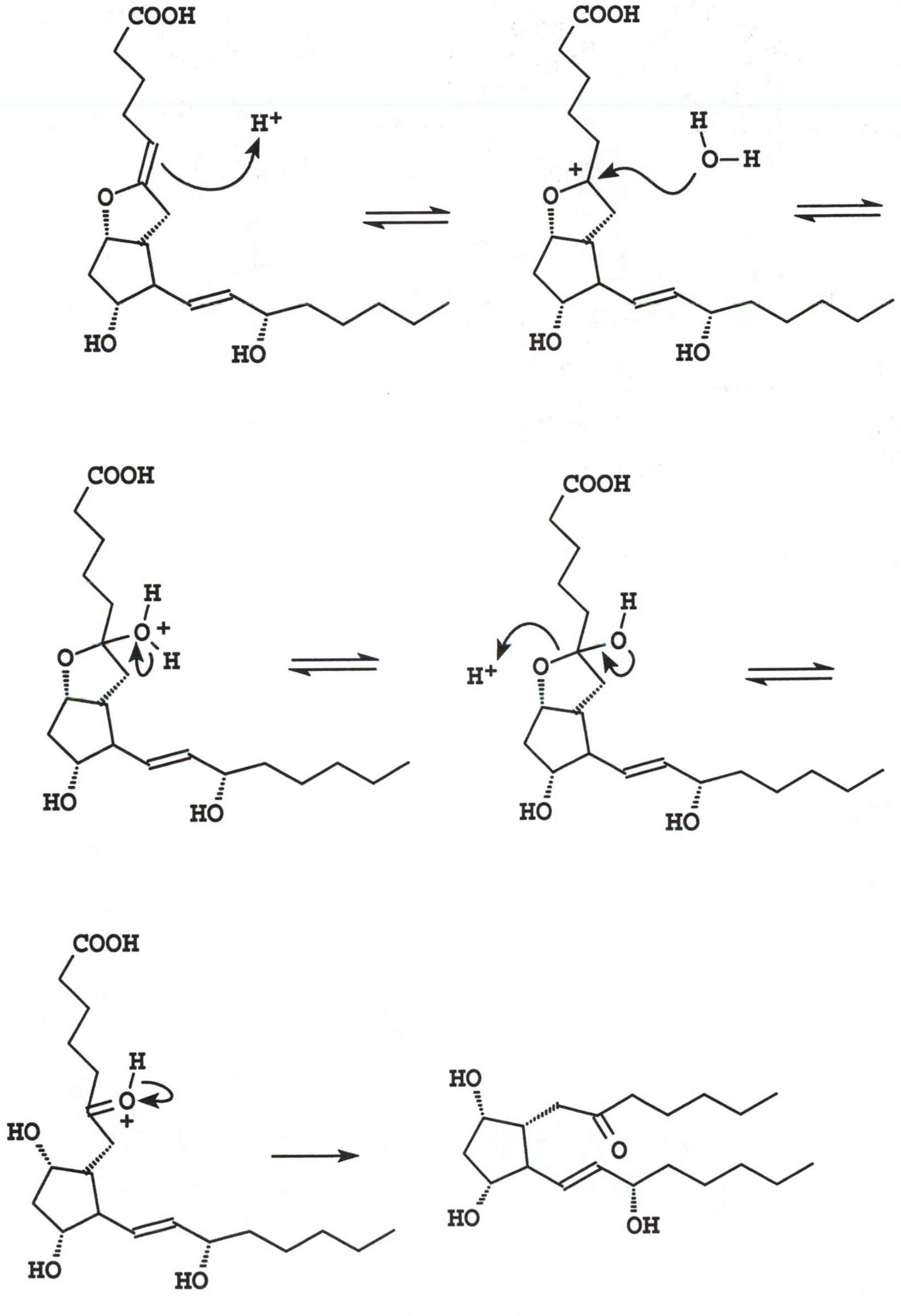
COOH
H+
H
O—H
+
COOH
COOH
H
O+
H
COOH
H
O
H+
COOH
H
O
+
HO
HO
HO
HO
HO
HO
HO
HO
HO
HO
HO
HO
O
OH

205)

256

206)

207)

257

208)

4

209)

258

210)

REFERENCES

Virtually all the problems cited in the workbook have been gleaned from the primary chemical literature. The original references to the problems cited in the workbook are given below. The numbering of the references corresponds to the numbering of the problems in the workbook.

1) ____

2) U. Azzena, S. Demartis, M. G. Fiori, G. Melloni, L. Pisano Tetrahedron Lett. 36, 8123 (1995).

3) U. Kraatz, M. N. Samimi, F. Korte Synthesis 430 (1977).

4) S. K. Taylor, S. A. May, J. A. Hopkins Tetrahedron Lett. 34, 1283 (1993).

5) H. Mazdiyasni, D. B. Konopacki, D. A. Dickman, T. M. Zydowsky Tetrahedron Lett. 34, 435 (1993).

6) J. Kaminska, M. A. Schwegler, A. J. Hoefnagel, H. van Bekkum Rec. Trav. Chim. Pays-Bas 111, 432 (1992).

7) J. Almena, F. Foubelo, M. Yus Tetrahedron 51, 3365 (1995).

8) J. Blank, W. Grosch, W. Eisenreich, A. Bacher, J. Firl Helv. Chim. Acta. 73, 1250 (1990).

9) C. Ferreri, M. Ambrosone, C. Chatgiliaglou Synth. Comm. 25, 3351 (1995).

10) S. N. Huckin, L. Weiler Can. J. Chem. 52, 2157 (1974).

11) J. Chen, M. T. Fletcher, W. Kitching Tetrahedron: Asymmetry 6, 967 (1995).

12) T. Cohen, F. Chen, T. Kulinski, Florio, S., Capriati, V. Tetrahedron Lett. 36, 4459 (1995).

13) ____

14) L. Djakovitch, J. Eames, R. V. H. Jones, S. MacIntyre, S. Warren Tetrahedron Lett. 36, 1723 (1995).

15) B. D. Brandes, E. N. Jacobsen J. Org. Chem. 59, 4378 (1994).

16) ____

17) B. Hagenbrunch, S. Hunig Chem. Ber. 116, 3884 (1983).

18) ____

19) R. B. Moffett, Org. Synth. Coll. Vol. 4, 238 (1963).

20) S. Masuda, T. Nakajima, S. Suga Bull. Chem. Soc. Jpn 56, 1089 (1983).

21) D. W. McCullough, T. Cohen Tetrahedron Lett. 29, 27 (1988).

22) A. Guy, J-P. Guette Synthesis 222 (1980).

23) Ref 7

24) R. S. Prasad, R. M. Roberts J. Org. Chem. 56, 2998 (1991).

25) ____

26) S. Kabagu, Y. Kojima J. Chem. Ed. 69, 420 (1992).

27) S. Kano, T. Ebata, S. Shibuya J. Chem. Soc. Perkins Trans I 2105 (1980).

28) J. Morris, D, G. Wishka Tetrahedron Lett. 29, 143 (1988).

29) ____

30) M. O. Fatope, J. I. Okogun J. Chem. Soc. I 1601 (1982).

31) S. Y. Dike, D. H. Ner, A. Umar Bioorg. Med. Chem. Lett. 1, 383 (1991).

32) M. G. Constantino, M. Beltrame, E. F. DeMedeiros, G-V. J. Da Silva Synth. Comm. 22, 2859 (1992).

33) Ref 7

34) B. M. Trost, M. J. Bogdanovich J. Am. Chem. Soc. 95, 5321 (1973).

33) Ref 7

34) B. M. Trost, M. J. Bogdanovich <u>J. Am. Chem. Soc.</u> **95**, 5321 (1973).

35) R. K. Boeckman, S. S. Ko <u>J. Am. Chem. Soc.</u> **102**, 7147 (1980).

36) M. R. Karim, P. Sampson <u>J. Org. Chem.</u> **55**, 598 (1990)

37) K. Koch, M. S. Biggers <u>J. Org. Chem.</u> **59**, 1216 (1994).

38) E. J. Corey, J. Das <u>Tetraherdon Lett.</u> **23**, 4217 (1982).

39) M. Karikomi, T. Yamazaki, T. Toda <u>Chem. Lett.</u> 1787 (1993).

40) N. Amlaiky, G. Leclerc, A. Carpy <u>J. Org. Chem.</u> **47**, 517 (1982).

41) (a) OS CV 6 704 (1988); (b) H. W. Pinnick, Y-H. Chang <u>J. Org. Chem.</u> **43**, 4662 (1978).

42) ____

43) S. Das, T. Karpha, M. Ghosal, D. Mukherjee <u>Tetrahedron Lett.</u> **33**, 1229 (1992).

44) F. Benedetti, F. Berti, A. Risaliti <u>Tetrahedron Lett.</u> **34**, 6443 (1993).

45) N. K. Sangwan, S. N. Rastogi <u>Chem. Ind. (London)</u>

46) N. Greeves, J. S. Torode <u>Synthesis</u> 1109 (1993)

47) ____

48) A. Gokhale, P. Schiess <u>Helv. Chim. Acta</u> **81**, 251 (1998); M. P. Cava, K. Muth <u>J. Am. Chem. Soc.</u> **82**, 652 (1960).

49) S-C. Kim, B-M. Kwon <u>Synthesis</u> 795 (1982).

50) A. P. Kozikowski, P. D. Stein <u>J. Am. Chem. Soc.</u> **104**, 4023 (1982).

51) S-C. Wong, S. Sasso, H. Jones, J. J. Kaminski <u>J. Med. Chem.</u> **27**, 20 (1984).

52) R. V. Stevens, A. W. M. Lee <u>Chem. Comm.</u> 102 (1982).

53) K. Skinnemoen, K. Undheim <u>Acta Chem. Scand. B</u> 295 (1980).

54) R. A. Bunce, C. J. Peeples, P. B. Jones <u>J. Org. Chem.</u> **57**, 1727 (1992).

55) G. G. Trigo, C. Avendano, E. Santos, H. N. Christensen, M. E. Handlongten <u>Can. J. Chem.</u> **58**, 2298 (1980).

56) H-J. Bertram, M. Guntert, H. Sommer, T. Thielmann, P. Werkhoff <u>J. Prakt. Chem.</u> **335**, 101 (1993).

57) K. Lovgren, A. Hedberg, J. J. G. Nilsson <u>J. Med. Chem.</u> **23**, 624 (1980).

58) ____

59) D. Zhang, G. L. Closs, D. D. Chung, J. R. Norris <u>J. Am. Chem. Soc.</u> **115**, 3670 (1993).

60) J-L. Gras, M. Bertrand <u>Tetrahedron Lett.</u> **47**, 4549 (1979).

61) C. J. Roxburgh <u>Tetrahedron</u> **47**, 10749 (1993).

62) A.Barco, S. Benetti, G. P. Pollini, P. G. Baraldi, C. Gandolfi <u>J. Org. Chem.</u> **45**, 4776 (1980).

63) J. E. Engelhart, J. R. McDivitt <u>J. Org. Chem.</u> **36**, 367 (1971).

64) K. Isido, S. Kurozomi, K. Utimoto <u>J. Org. Chem.</u> **34**, 2661 (1969).

65) K. Gothelf, I. Thomsen, K. B. G. Torsell <u>Acta Chem. Scand.</u> **46**, 494 (1992).

66) F. Loftus <u>Synth. Comm.</u> **10**, 59 (1980).

67) D. P. Curran <u>Tetrahedron Lett.</u> **24**, 3443 (1983).

68) R. C. Cookson, P. S. Ray <u>Tetrahedron Lett.</u> **23**, 3521 (1982).

69) ____

70) (a) J. Babbler, S. A. Schlidt <u>Tetrahedron Lett.</u> **33**, 7697 (1992); (b) M. Rosenberger, W. Jackson, G. Saucy <u>Helv. Chim. Acta</u> **63**, 1665 (1980).

71) P. A. Zoretic, F. Barcelos, J. Jardin, C. Blakta <u>J. Org. Chem.</u> **45**, 810 (1980).

72) P. D. Gardner, G. R. Hayenes, R. L. Brandon J. Org. Chem. **22**, 1206 (1957).

73) J. D. Albright, D. B. Moran, W. B. Wright, J. B. Collins, B. Beer, A. S. Lippa, E. N. Greenblatt J. Med. Chem. **24,** 592 (1981)

74) R. A Bunce, M. J. Bennett Synth. Comm. (1993) [in press]

75) J. Morris, D. G. Wishka Tetrahedron Lett. **29**, 143 (1988).

76) G. Rousseau Tetrahedron **51**, 2777 (1995).

77) N. Amlaiky, G. Leclerc Synthesis 426 (1982)

78) ____

79) D. B. Reitz J. Org. Chem. **44**, 4707 (1979).

80) F. Busque, Cid, P. de March, M. Figueredo, J. Font Hetereocycles **40**, 387 (1995).

81) Z-F. Xie, H. Suemune, K. Sakai Synth. Comm. **19**, 901 (1989).

82) M. Girard, D. B. Moir, J. W. ApSimon Can. J. Chem. **65**, 189 (1987).

83) J. N. Marx, G. Minaskanian J. Org. Chem. **47**, 3306 (1982).

84) K. Sato, S. Inoue, K. Ozawa, T. Kobayashi, T. Ota, M. Tazaki J. Chem. Soc. Perkin Trans I 1753 (1987)

85) ____

86) F. J. Urban Tetrahedron: Asymmetry **5,** 211 (1994).

87) J. M. Nicholson, I. O. Edafiogho, J. A. Moore, V. A. Farrar, K. R. Scott J. Pharmaceut. Sci. **83**, 76 (1994).

88) D. Seebach, T. Hoffmann, F. N. M. Kuhnle, J. N. Kinkel, M. Schulte Helv. Chim. Acta **78**, 1525 (1995).

89) D. H. Kim J. Het. Chem. **12**, 1323 (1975)

90) J. Jernow, W. Tautz, P. Rosen, J. F. Blount J. Org. Chem. **44**, 4212 (1979).

91) H. Hikino, P. De Mayo J. Am. Chem. Soc. **86**, 3582 (1964).

92) ____

93) Y. Fall, L. Santana, M. Teijeira, E. Uriarte Heterocycles **41**, 647 (1996).

94) F. J. McEvoy, J. D. Albright J. Org. Chem. **44**, 4597 (1979).

95) G. N. Walker and D. Alkalay J. Org. Chem. **36**, 492 (1971).

96) S. Suzuki, H. Stammer Bioorg. Chem. **15**, 43 (1987).

97) Ref 87

98) P. G. Baraldi, A. Barco, S. Benetti, G. P. Pollini, V. Zanirato Tetrahedron Lett. 4291 (1984).

99) C-P. Chang, L-F. Hsu, N-C. Chang J. Org. Chem. **59**, 1898 (1994).

100) ____

101) J. Almena, F. Foubelo, M. Yus J. Org. Chem. **61**, 1859 (1996).

102) H. A. Stephani, N. petragnani, C. J. Valduga, C. A. Brandt Tetrahedron Lett. **38**, 4977 (1997).

103) L. G. Mueller, R. G. Lawton J. Org. Chem. **44**, 4741 (1979).

104) J. P. Wasacz, V. G. Badding J. Chem. Ed. **59**, 695 (1982).

105) T. Hiyama, M. Shinoda, H. Nozaki J. Am. Chem. Soc. **101**, 1599 (1979).

106) B. R. Davis, M. G. Hinds Austr. J. Chem. **50**, 309 (1997).

107) T. Cohen, M. Bhupathy, J. R. Matz J. Am. Chem. Soc. **105,** 520 (1993).

108) C. D. Apostolopoulos, S. Spyroudis, P. Tarantili J. Het. Chem. **33**, 703 (1996).

109) M. Abdul-Aziz, J. V. Auping, M. A. Meador J. Org. Chem. **60**, 1305 (1995)

110) H. Sard, F. Shawcross J. Het. Chem. **32**, 1393 (1996).

111) E. L. Williams Synth. Comm. **22**, 1017 (1992).

112) E. Valencia, A. J. Preyer, M. Shamma, V. Fajardo Tetrahedron Lett. 599 (1984).

113) D. J. Hart, A. Kim, R., Krishnamurthy, G. H. Merriman, A-M. Waltos Tetrahedron 48, 8179 (1992).

114) E. L. Williams Synth. Comm. 22, 1017 (1992)   [CHECK]

115) G. Stefancich, M. Artico, S. Massa, S. Vomero J. Het. Chem. 16, 1443 (1979).

116) K. C. Nicolaou, S. P. Seitz, W. J. Sipio, J. F. Blount J. Am. Chem. Soc. 101, 3884 (1979).

117) Ref 4

118)   J. P. Dittami, F. Xu, H. Qi, M. W. Martin, J. Bordner, J. Kiplinger, P. Reiche, R. Tetrahedron Lett. 36, 4197 (1995).

119) B. M. Branan, L. A. Paquette, J. Am. Chem. Soc. 116, 7658 (1994).

120)   K. Gollnick, G. Schade, A. F. Cameron, C. Hannaway, J. S. Roberts, J. M. Robertson Chem. Comm. 248 (1970).

121) X. Shi, S. Miller J. Org. Chem. 58, 2907 (1993).

122) K. E. Andersen, C. Braestrup, F. C. Gronwald, A. S. Jorgensen, E. B. Nielsen, U. Sonnewald, P. O. Sorensen, P. D. Suzdak, L. J. S. Knutsen J. Med. Chem. 36, 1716 (1996).

123) G. Descotes, D. Missos Synthesis 149 (1971).

124) W. Y. Lee, C. H. Park, E. H. Kim J. Org. Chem. 59, 4495 (1994).

125) J. T. Negri, R. D. Rogers, L. A. Paquette J. Am. Chem. Soc. 113, 5073 (1991).

126) K. D. Raner, C. R. Strauss, F. Vyskoc, L. Mokbel J. Org. Chem. 58, 950 (1993).

127) T. Aida, R. Legault, D. Dugat, T. Durst Tetrahedron Lett. 4993 (1979).

128) O. Aleksiuk, S. Cohen, S. E. Biali J. Am. Chem. Soc. 117, 9649 (1995).

129) P. Galatsis, J. J. Manwell Tetrahedron Lett. 36, 8179 (1995).

130) A. Nath, J. Mal, R. V. Venkateswaran Chem. Comm. 1374 (1993).

131) P. Cannone, D. Belanger, G. Lemay Synthesis 301 (1980)

132) X-F. Ren, E. Turos, C. H. Lake, M. R. Churchill J. Org. Chem. 60, 6468 (1995).

133) Ref 88

134) ____

135) S. M. Allin, C. C. Hodkinson, N. Taj Synlett. 781 (1996).

136) S. Takano Chem. Lett. 359 (1989).

137) P. Martin, T. Winkler Helv. Chim. Acta 77, 100 (1994).

138) I. Vlattas, I. T. Harrison, L. Tokes, J. H. Fried, A. D. Cross J. Org. Chem. 33, 4176 (1968).

139) P. Kocienski, C. Yates Tetrahedron Lett. 3905 (1983).

140) L. R. Smith, H. J. Williams J. Chem. Ed. 56, 696 (1981).

141) H. Mayr, J-P. Dau-Schmidt Chem. Ber. 127, 213 (1994).

142) C. W. Ong, C. M. Chen, S. S. Juang J. Org. Chem. 59, 7915 (1994).

143) A. Padwa, S. S. Murphee, P. E. Yeske J. Org. Chem. 55, 4241 (1990).

144) M. T. Crimmins, D. M. Bankaitis Tetrahedron Lett. 4551 (1983).

145) M-C. Carre, B. Gregoire, P. Caubere J. Org. Chem. 49, 2050 (1984).

146) R. D. Crouch, T. D. Nelson J. Chem. Ed. 72, A6 (1995).-

147) Ref 62

148) R. P. Nelson, J. M. McEwen, R. G. Lawton J. Org. Chem. 34, 1225 (1969).

149) P. G. Baraldi, A. Barco, S. Benetti, F. Moroder, G. P. Pollini, D. Simoni, V. Zanirato Chem. Comm. 1265 (1982).

150) H. Yoshimura, M. Nagai, S. Hibi, K. Kikuchi, S. Abe, T. Hida, S. Higashi, I. Hishinuma, T. Yamanaka <u>J. Med. Chem.</u> **38**, 3163 (1995).

151) T. Kawano, T. Ogawa, S. M. Islam, I. Ueda <u>Tetrahedron Lett.</u> **36**, 7685 (1995).

152) M. S. South <u>J. Het. Chem.</u> **28**, 1013 (1991).

153) S. V. Ley, R. Leslie, P. D. Tiffin, M. Woods <u>Tetrahedron Lett.</u> **33**, 4767 (1992).

154) P. Bakuzis, M. L. F. Bakuzis, T. F. Weingartner <u>Tetrahedron Lett.</u> 2371 (1978).

155) J. P. Dickens, G. J. Ellames, N. J. Hare, K. R. Lawson, W. R. McKay, A. P. Metters, D. L. Myers, A. M. S. Pope, R. M. Upton <u>J. Med. Chem.</u> **34**, 2356 (1991).

156) ____

157) D. J. Dunham, R. G. Lawton <u>J. Am. Chem. Soc.</u> **93**, 2074 (1971).

158) R. J. K. Taylor, S. M. Turner, D. C. Horwell, O. W. Howarth, M. F. Mahon, K. C. Mollon <u>J. Chem. Soc. Perkin Trans I</u> 2145 (1990).

159) W. C. Vincek, C. S. Aldrich, R. T. Borchardt, G. L. Grunewald <u>J. Med. Chem.</u> **24**, 7 (1981).

160) ____

161) A. V. Rama Rao, M. N. Deshmukh, M. Kamalam <u>Tetrahedron</u> **37**, 227 (1981).

162) A. Marfat, P. Helquist <u>Tetrahedron Lett.</u> 4217 (1978).

163) P. F. Schaltz <u>J. Chem. Educ.</u> 468 (?)

164) P. T. Lansbury, A. K. Serelis <u>Tetrahedron Lett.</u> 1909 (1978).

165) R. B. Gamill, S. A. Nash, S. A. Mizsak <u>Tetrahedron Lett.</u> **24**, 3435 (1983).

166) E. Campaigne, R. A. Forsh <u>J. Org. Chem.</u> **43**, 1044 (1978).

167) ____

168) (a) M. Tramontini <u>Synthesis</u> 703 (1973); (b) M. Tramiontini, L. Angiolini <u>Tetrahedron</u> **46**, 1791 (1990).

169) V. Reutrakul, W. Kanghae <u>Tetrahedron Lett.</u> 1225 (1977).

170) ____

171) ____

172) R. M. Scarborough, B. H. Troder, A. B. Smith <u>J. Am. Chem. Soc.</u> **102**, 3904 (1980).

173) R. Lowe, P. Jeske <u>Annalen</u> 549 (1987).

174) Ref 165

175) S. S. Khatana, D. H. Boschelli, J. B. Kramer, D. T. Connor, H. Barth, P. Stoss <u>J. Org. Chem.</u> **61**, 6060 (1996).

176) ____

177) W. S. Johnson and G. H. Daub <u>Org. React.</u> **6**, 1 (1951)

178) R. Queignec, B. Kirschleger, F. Lambert, M. Aboutaj <u>Synth. Comm.</u> **18**, 1213 (1988)

179) D. F. Taber, J. C. Amedio, K-Y. Jung <u>J. Org. Chem.</u> **52**, 5621 (1987).

180) J. Bagli, T. Bogri, K. Voith <u>J. Med. Chem.</u> **27**, 875 (1984).

181) Ref 31

182) J. Arct, E. Jakubska, O. Olszueska <u>Synth. Comm.</u> **8**, 143 (1978).

183) F. Van Hove, S. Vanwtswinkel, J. Marchand-Brynaert, J. Fastrez <u>Tetrahedron Lett.</u> **36**, 9313 (1995).

184) A. Armstrong <u>Encyclopedia of Reagents for Organic Synthesis</u> 2, 1006 (1995).

185) M. Sato, J. Sakaki, Y. Sugita, S. Yasuda, H. Sakoda, C. Kaneko <u>Tetrahedron</u> **47**, 5689 (1991).

186) ____

187) R. Zehnter, H. Gerlach Liebigs Ann. 2209 (1995).

188) E. P. Boden, G. E. Keck J. Org. Chem. **50**, 2394 (1985).

189) J. V. Cooney J. Het. Chem. **20**, 823 (1983).

190) T. M. Bargar, C. M. Riley Synth. Comm. **10**, 479 (1980).

191) J. J. Tegeler, H. H. Ong, J. A. Profitt J. Het. Chem. **20**, 867 (1983).

192) R. Bayles, M. C. Johnson, R. F. Maisey, R. W. Turner Synthesis

193) R. K. Boeckman, S. S. Ko J. Am. Chem. Soc. **102**, 7147 (1980).

194) F. M. Hauser, R. P. Rhee J. Org. Chem. **43**, 178 (1978).

195) S. N. Huckin, L. Weiler J. Am. Chem. Soc. **96**, 1082 (1974).

196) L. A. Paquette, J. P. Freeman J. Org. Chem. **35**, 2249 (1970).

197) S. Raina, V. K. Singh Tetrahedron **51**, 2467 (1995).

198) D. L. Hughes Org. React. **42**, 335 (1992).

199) A. Armstrong, J. Brackenridge, R. F. W. Jackson, J. M. Kirk Tetrahedron Lett. **29**, 2483 (1988).

200) J. P. Muxworthy, J. A. Wilkinson, G. Procter Tetrahedron Lett. **36**, 7539 (1995).

201) M. Gerlitz, G. Udvarnoki, J. Rohr Angew. Chem. Int. Ed. **34**, 1617 (1995).

202) ____

203) N. P. Jensen, J. J. Friedman, H. Kropp, F. M. Kahan J. Med. Chem. **23**, 6 (1980).

204) ____

205) S. J. Behroozi, W. Kim, K. S. Gates J. Org. Chem. **60**, 3964 (1995).

206) ____

207) ____

208) D. W. Norbeck, W. Rosenbrook, J. B. Kramer, D. J. Grampovnik, P. A. Lartey J. Med. Chem. **32**, 625 (1989).

209) F. R. Kinder, K. W. Bair J. Org. Chem. **59,** 6965 (1994).

210) Y-Z. Shu, Q. ye, J. M. Kolb, S. Huang, J. A. Veitch, S. E. Lowe, S. P. Manly J. Nat. Prod. **60**, 529 (1997).